DESIGNING FOR NOISE AND NATURAL VENTILATION

DESIGNING FOR NOISE AND NATURAL VENTILATION

A PRACTICAL GUIDE FOR NON-ACOUSTICIANS

By

J. T. Cooke

Media, London

Copyright

Title: Designing for Noise & Natural Ventilation; A Practical Guide for Non-Acousticians
Author: J. T. Cooke

Published by

Media

First edition published in 2021
Contact: www.kuu-media.com

For information about student discounts and special quantity discounts please contact Kuu Media.

A CIP catalogue record for this book is available from the British Library.

Includes bibliographical references, index and appendix.

ISBN: 978-1-7398692-0-5 (pbk.), 978-1-7398692-1-2 (ebook), 978-1-7398692-2-9 (hbk)

Subjects: Design, Urban Planning, Architecture, Urban Design, Acoustics. Dewey No: 720, 690.

Cover photo, 110 Freeway in downtown Los Angeles: by Nicole De Khors from Burst. Rear page photo by JT Cooke

DEDICATION

For Kasper, and dedicated to my family.

In memory of Jaime Kleinert

ACKNOWLEDGEMENTS & THANKS

No person ever works in a vacuum, and the author would like to acknowledge and thank all those who helped him and taught him over the years.

Thanks to all those who helped me and gave permission to allow their images to be reproduced in this book; the list is a long one, and forgive me if I forget anyone; including but not limited to:

Organisation	*Person*
	Anne Finn
	Adam Haddow
	Brian Cox
	Bridget Smyth
	Cindy Ch'ng
	Damian Ennis
	David Reynolds
	Jesse McNicoll
	Kate Luckraft
	Kelley Shaw
	Matt McNeil
	Paula Hellén
	Peter John Cantrell
	Sacha Coles
	Tony Smith

Organisation	*Person*
Artnovion	
Bruag AG	
Department of Environment	
ECE Architecture	
Edinburgh Napier University	Richard Mackenzie
Eilenburger Fenstertechnik GmbH & Co. KG	Jens Schimmel & Wolf von Trotha
Environa Studios	Tone Wheeler
Environmental Protection Department; Government of Hong Kong, Special Administrative Region	Maurice Yeung
Force Technology	Lars Sommer Søndergaard
H+N+S Landscape Architects	Jutta Raith
Krieger Architekten	Peter Krieger
Loosen, Rüschoff + Winkler, Architekten und Stadtplaner	Karin Loosen
Loosen, Rüschoff + Winkler, Architekten und Stadtplaner	Kilian Jonak
Publications office of the European Union	
The (Danish) State Building Research Institute at Aalborg University Copenhagen	
The Ministry of Food of Denmark; Environment Protection Agency	
The Sydney & NSW Design Community	
The Sydney & NSW Planning Community	
U.S. Federal Highway Administration; Office of Planning, Environment, and Realty	Aileen Varela-Margolles
University of Cambridge	Alex Wynick
World Health Organization	
Aumann Katzsch Architekten GmbH	Arne Aumann

HOW TO USE THIS BOOK

In our post-pandemic world, we have all become adept at using QR codes on our phones. This book includes links as QR codes or hyperlinks to additional free resources and external websites for the reader's convenience. The book includes QR codes that the reader can scan with their phone or tablet camera to take them to a map location or website.

Understanding the context of a building is crucially important to know why a specific noise solution is appropriate. Example projects contain hyperlinks and QR codes that take the reader to the project's location on OpenStreetMap, Google Maps, or Bing Maps for convenience. If the reader wishes, they can take time to explore the broader context of examples through OpenStreetMap, Google Maps, or Bing Maps aerial and understand the position of the project in the settlement. Understanding the wider settlement context through exploring maps may help the reader understand why the chosen noise solution addresses the specific context.

As time progresses, I hope to update the book with new solutions and innovations, so please subscribe for updates and join our mailing list. Please scan the QR codes below using your phone to take you to the Kuu Media website and YouTube Channel, where you can subscribe for videos and additional free information.

For those who like more visual content, related video tutorials of designing for noise are available on the Kuu Media website.

Kuu Media website

YouTube Channel

Other sections of the book referred to in the body text and items with further definition and explanation in the *glossary of terms and definitions* are in *italics*

CONTENTS

FOREWORD

Brian Cox CIEH, MSc Acoustics

Noise is unwanted sound. A pleasant sound like birdsong, a babbling brook or your favourite music can enhance your acoustic environment, but an undesirable sound or noise can diminish it. The potential adverse health impacts of noise have been known for a long time, but more recently, the understanding of the mechanism behind those negative impacts is better understood.

As early humans, we were programmed to continually listen for sounds which could mean danger for ourselves or those around us. Those noises that could relate to danger cause a release of adrenalin aimed to prepare the body for action. In the modern world for the majority of the time, there is no longer a need to react, so the adrenalin remains unutilised and starts to work against the body as cortisol. In the urban environment, such noises occur multiple times every day and night, and over long periods of time, and this causes lasting damage to the human body. Even when asleep, our brains are listening, with unknown or unwanted sounds triggering a response causing the brain to arouse out of deep restorative sleep to lighter sleep. If repeated regularly, this results in a multitude of potential adverse health impacts.

Unfortunately, this primitive defence mechanism designed to protect us is now working against us. We have ultimately become victims of our own evolution. It is, therefore, imperative to protect people from the worst aspects of noise. Most developed nations have strong laws to deal with behavioural noise from neighbours, but it is the environmental noise that is difficult to control, especially in the urban environment. This is why protecting residents' internal noise environment is so important.

In my role of public health and noise enforcement in Local Government, I am tired of receiving plans for poorly thought-out housing schemes where acoustics and noise control are always an afterthought and usually consist of enhanced double-glazing and mechanical ventilation at best. Working in a compact urban environment, most of the simple mitigation tools of separation and barriers were not available.

However, my thinking was expanded by the author's in-depth knowledge of design, which opened my eyes to a multitude of options. This knowledge has now been channelled into a single document Designing for Noise and Natural Ventilation: a practical guide for non-acousticians, which should be a staple for designers, planners and Environmental Health Practitioners across the world. It demonstrates the effectiveness of using simple design features and changes of building layout, so ensuring an acceptable internal noise environment, while still maintaining a good flow of natural ventilation to aid cooling and afford better air quality and, in turn, protecting the residents from the adverse impacts of noise while maintaining a pleasant living environment. It clearly shows that with a little thought, good acoustic design can be done and more importantly, it should be done.

PREFACE

I spent part of my life growing up next to a busy road in London, near Heathrow Airport, and I can remember on the warm summer nights having to choose between sweltering in the heat or leaving the windows open to try to keep cool. I remember being exhausted at school the following day because I had not gotten enough sleep due to the noise and heat. Later in my life, I once again lived in a noisy area, but this time in a brand-new apartment building, and once again, I had to choose between lack-of-sleep, because of noise, or overheating. Unfortunately, not much had improved in twenty years.

Noise exposure is a pervasive public health issue. Still, there is little understanding of the solutions for noise that are available, nor is there an understanding of where to deploy them in our settlements, and where specific noise solutions may be more appropriate. Many of the noise mitigation solutions proposed to the author while working in industry have been more suitable for a suburban, or a rural context or designers have defaulted to specifying air-conditioning. I hope that this book will also provoke all involved in the planning and design of human habitat that you are not 'just' planning or designing 'flats', or 'apartments' you are creating homes for people. The buildings made will be the housing stock, and dwellings, offices, and schools of your settlements for decades, and they will form the people and place that will go on to be your settlement and community. They are too important to get wrong.

The problems of natural ventilation and noise are often dismissed as a luxury, 'nice to have', but unnecessary, or overlooked because of the primacy of the visual sense in design. In this time of a Covid-19 pandemic, the need for a plentiful supply of fresh, clean, air to indoor spaces is even more critical, and recycled air-conditioned air, circulating the same stale air is something to be avoided. Energy hungry air-conditioning is also unsustainable, expensive to run, and passes the cost of a poor design onto the resident.

The lack of knowledge among some designers, planners, developers, and peoples representatives about how to design for noisy areas while providing natural ventilation, and what a good design looks like, has surprised me. I wrote this holistic, settlement focussed, practical guidebook to fill some gaps in the knowledge of several fields. I have written the book in a way that is hopefully, easy to understand, and accessible for a broad professional audience, and explanatory videos are available. I hope this guidebook highlights an often-overlooked issue in design and planning and publicises the many simple and practical solutions available.

I first began the process of researching noise and sound in human settlements at university, and one of the first projects I completed was an audible experience map of a walk through London set beside a visual map of the same walk. In covering my eyes and experiencing the city through the sound it makes, forced me to focus on how much the noises in our settlements affect people.

Over the years, I produced and came across different techniques and strategies for designing and planning for noise in our settlements, and I have compiled them in this book. I began researching better ways to design for noise and alleviate the severe implications for public health about six years ago. I have been writing this book for the past five years and faced several challenges along the way, but finally, it is complete, and I hope it is of some use.

I went through several iterations of the title. I considered Designing & Planning for Natural Ventilation without Noise, Designing for Natural Ventilation without Noise, and Designing & Planning for Noisy Contexts and Natural Ventilation. The first title probably reflects the contents more accurately, but the current title was thought to be more catchy.

Being continually confronted with the low standard of the design and execution of many apartment buildings, including the one I lived in, and a conviction that there must be a better way were the catalysts for writing this book. It frustrated me that strategic planning, urban design and planning assessment didn't take the issue seriously, and we could do more to help solve this real problem and create good quality places for people to live.

For the love of the city and its people.

INTRODUCTION

Noise in human habitat is often dismissed as inconsequential, but it is a pervasive problem.

The evidence of the detrimental effect on human health is clear. The World Health Organisation (WHO) has published several reports providing evidence of the damaging impact of noise on human health. WHO publications including; the Burden of Disease from Environmental Noise: Quantification of Healthy Life Years Lost in Europe, in 2011, and the Environmental Noise Guidelines for the European Region, in 2018, provide proof of the detrimental effect noise is having on Public Health. The techniques in the book seek to give control over noise ingress back to the occupants and improve living conditions. This book offers a practical guide for non-acousticians to design and plan to reduce noise exposure of our population, and provides practical techniques and examples to illustrates many design solutions to mitigate noise, while concurrently also providing natural ventilation.

The United Nations estimates that 68% of the world population will live in urban areas by 2050, increasing the likelihood of noise exposure to an increasing proportion of our society (United Nations, 2018). The growing number of the world's population living in dense urban-centres will require innovative passive acoustic design solutions to provide natural ventilation without noise, and reduce human settlements' energy consumption. Currently, where human habitation is proximate to noise sources, an acoustically sealed building with energy-consuming air-conditioning is overwhelmingly the default solution, but this is unsustainable. In March 2018, the United Nations Secretary-General António Guterres called climate change, '...the most systemic threat to humankind...' [6] (United Nations, 2018, p. 1). Built-environment professionals cannot continue to specify energy-consuming air-conditioning to solve ventilation, overheating, and noise issues. Passive methods of providing natural ventilation for cooling, while maintaining acoustic privacy are critical and provide a design challenge. Research is underway into noise-cancelling technology; however, this technology requires energy and can only cancel certain frequencies, and this book focuses exclusively on solutions that do not require power.

Natural ventilation for cooling is a passive cooling strategy and is beneficial because it does not consume energy, unlike air-conditioning. In providing natural ventilation for respiration and cooling buildings, design solutions need to mitigate noise, particularly at night-time, to prevent sleep disturbance. Because of the need for natural ventilation, the book focusses on air-borne noise entering buildings through ventilation openings and only touches on noise transfer through building fabric.

Much noise affected development is in areas previously deemed not suitable as human habitat in our settlements such as; industrial areas, or near roads. New roads and infrastructure can also cause noise problems and blight areas of our settlements that had been suitable habitat. Since the late 1980s, strategic planning has sought to consolidate new development in existing settlements in the interests of sustainability, however; there has not been an investment in providing the tools and knowledge required to design and plan in these new territories of our settlements. The need to design for the specific characteristics and issues of a context is not new, and it is even noted in De Architectura by Vitruvius, circa 30BCE. A better understanding of the structure of settlements and the passive solutions available is required. This book illustrates practical solutions to noise and a better understanding of where in human settlements, they are appropriate, to provide good human habitat and excellent design. The book deliberately frames all discussion within the structure of human settlements and habitat. It puts people and human settlements as the starting point and common denominator to relate all design to people, and provide a way to organise noise issues and solutions. Hopefully, conceptualising human settlements, as human habitat will provoke all professions to remember that we aim to create better habitat for humans. Focussing on creating an excellent human habitat can give the disparate professional fields of; governance, law, public health, planning, engineering, transport planning, urban design and architecture a common reference point and aim. The Settlement and Habitation Transect is suggested as a tool to inform communities and help coordinate the approach of all professionals involved in the creation and management of settlements. The Settlement and Habitation Transact is used to organise where in our settlements particular noise reduction interventions are more likely.

The book is written for a multi-disciplinary professional audience and to cross the boundaries of professional fields. Because the book crosses disciplines, the reader may find they are reading subject matter that wouldn't typically be in a discipline-specific guide. The book draws on acoustics, design, urban design, urban planning, public health, and math. The cross-disciplinary approach aims to expose all disciplines to holistic information and help a rounded understanding of the practicalities of designing for noise. Sometimes we are so focussed on the trees of our profession we don't see the woods.

The first section of the book discusses; the physics of sound, the history and formation of human settlements and noise in human settlements, the general form of human settlements and the likely noise sources. The first section also discusses the public health issues created by noise and outlines how urban planning has tried to control the problem of noise in our settlements.

The second section of the book focuses on noise control solutions. Different solutions may be more appropriate depending on the project's location in a settlement and the development type. The book discusses the acoustic properties of materials, and how shape and form affect acoustic performance. The book divides the potential noise solutions into their likely practical applicable location in a settlement. The book discusses the various noise reduction techniques and categorises them in relation to the design process and place in a settlement. The solutions range from a building component, to site layout and urban design solutions. A range of solutions is illustrated, including:

Small interventions that are more likely for existing building stock and the centre of a settlement.

More significant interventions such as the layout of buildings on a site and solutions that are more likely for larger developments. Design solutions that are more likely for new buildings and toward the periphery of settlements. Regional-scale solutions, including the separation of noise sources from settlements.

CHAPTER 1

THE PHENOMENA OF SOUND

What is Sound?

It is essential to understand the role of human perception and the interaction between the various forms of energy in our habitat and our senses to understand the phenomenon of sound. Regarding the sense of hearing, it may be helpful to forget the word sound for a moment and think instead of energy.

1.1 HUMAN PERCEPTION OF OUR HABITAT

How we perceive energy

Our human brains interpret energy stimuli from our habitat through our senses and determine how we respond to them. Humans have several senses, and we can simplify them into five basic human senses that perceive our habitat. The five basic senses are; touch, sight, hearing, smell and taste, and they all perceive energy and categorise it. The sense of touch can interpret energy in our habitat as pressure and hot and cold on our skin. Sight interprets energy as light and the absence of light in our eyes. Our sense of hearing interprets the energy from small fluctuations in air pressure as sound and the lack of sound in our ears. The senses of smell and taste are similar because they interpret chemical energy in our habitat and categorise it.

Our senses are energy receivers for our brain, and they work together to interpret our habitat. The sense of touch can also perceive energy in the form of pressure waves as vibrations, which our ears can also sense. Our eyes perceive light and darkness but cannot perceive the sun's warmth nor the cool of the night, and for this, we use our sense of touch. Sound is energy in the form of a pressure wave passing through a medium to you, the receiver. It is essential to understand that you, as the receiver of sound energy, is critical in interpreting that sound energy as noise, music, speech etc. and if that sound energy could signal a potential threat.

1.2 WHAT IS NOISE?

Noise is subjective

Noise has been a part of human settlements for thousands of years, and the decision, when a sound is loud enough and unpleasant enough that it causes a person disturbance, is subjective. Noise is just unwanted sound energy and it may be unwanted because it is loud, or unpleasant, and it is the listener who decides if sound energy is noise or music. Energy has many forms, and to remove unwanted sound, you are trying to remove the energy by deflecting it or convert it to another form of energy, such as heat. A helpful metaphor for understanding the difference between sound and noise is weeds in a garden. Weeds are just unwanted plants, and noise is just unwanted sound. One gardener's weed may be another's prize flower, and in the same way, what is noise to one person may be pleasant music to another. What constitutes noise varies from person to person. Because of this personal preference, the issue is really about control over whether one is subjected to the sound waves one perceives as noise. It is about control and being able to control one's environment and habitat.

1.3 THE PHYSICS OF SOUND

Understanding sound

For humans to perceive sound, it must have a medium to pass through, which is part of one's habitat. In the case of whales, the medium is usually water, and for humans, that medium is typically the air. It is important to remember that sound is energy passing through the air, and changes to the air will affect how the sound behaves. The effect of the sound energy on the molecules in the air will be different if the air is warm or cold, humid or dry, and at higher or lower atmospheric pressure. Measuring air temperature and air pressure are different ways of measuring the energy and level of excitedness of molecules in the air. Where the air is the medium, pressure and temperature are linked. Temperature is a measure of the movement, or the excitement of molecules and the higher the temperature, the greater the movement. The pressure is a measure of the force the molecules exert on other matter, or you, and the higher the pressure, the more the molecules exert. High pressure and high temperature go together in the atmosphere, and the warmer it is, the more the molecules move and the more pressure they exert.

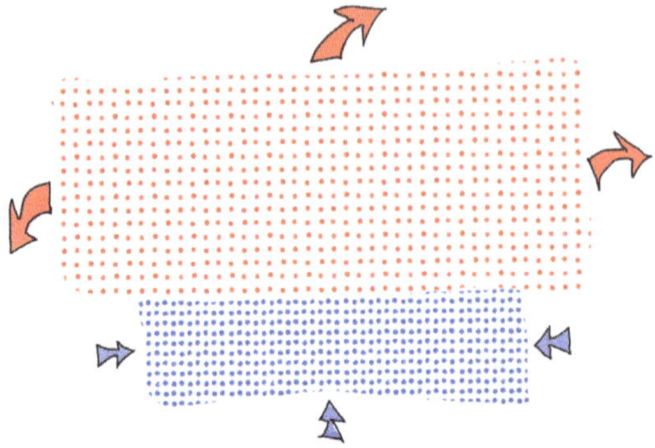

FIGURE 1 - UNDERSTANDING AIR AND PRESSURE; A DIAGRAM ILLUSTRATES THE SAME NUMBER OF AIR MOLECULES WHEN COLD AND WHEN WARM, AND THEY OCCUPY DIFFERENT AMOUNTS OF SPACE; BY J. T. COOKE; 2020 © J.T.COOKE. *NOTE: WHEN AIR IS COOLED THE MOLECULES ARE CLOSER TOGETHER AND WHEN WARM THEY ARE FURTHER APART. WHEN AIR MOLECULES GO FROM COOL TO WARM, THEY SPACE OUT MORE WHICH EXERTS AN OUTWARD PRESSURE, WHEN THEY GO FROM WARM TO COOL THE MOVE CLOSER TOGETHER, WHICH CAUSES A VACUUM DRAWING OTHER AIR MOLECULES TOWARD THE VACUUM, A BIT LIKE AN INWARD PRESSURE.THE METAPHOR OF PEOPLE ON A DANCEFLOOR MAY BE HELPFUL WHEN THINKING OF THE EFFECT OF HEAT ON AIR MOLECULES AND AIR PRESSURE. THE COOL MOLECULES ARE LIKE A SLOW DANCE, AND THE MOLECULES MOVE SLOWER AND ARE CLOSER TOGETHER. THE WARM MOLECULES ARE LIKE A CÈILIDH OR A SWING DANCE, AND MOLECULES MOVE FASTER AND NEED MORE SPACE TO MOVE.*

The speed that sound travels through the air is affected by the temperature of the air and humidity levels. You may have experienced this phenomenon when leaving bedroom windows open on hot, still summer nights if you have noticed hearing the sound of distant trains that you cannot usually hear.

The sound energy passing through the air causes sound pressure variances in the underlying atmospheric pressure and this is called Acoustic Pressure, which is measured in Pascals (Pa) and some explanations of acoustic terminology are in Table 1. Acoustic Pressure and Atmospheric Pressure are both measures of forms of energy in the medium of air, and because atmospheric pressure is ever present in our habitat human hearing has evolved to largely ignore the underlying atmospheric pressure and focus on the small fluctuations in the pressure caused by sound energy. Picture yourself at a beach, in a fairly calm sea with the waves bobbing up and down and lots of other swimmers splashing in the water. The waves of the sea bobbing up and down are like the atmospheric pressure and the smaller waves made by swimmers splashing are like the acoustic pressure waves and operate in conjunction with the sea waves, see Figure 4. Just like the smaller waves made by splashing swimmers, with sound pressure waves, the molecules in the air are not at a consistent level of movement, they vary, and this variance is called Frequency and is measured in Hertz (Hz). It may be helpful to watch the video at the QR code link for further explanation.

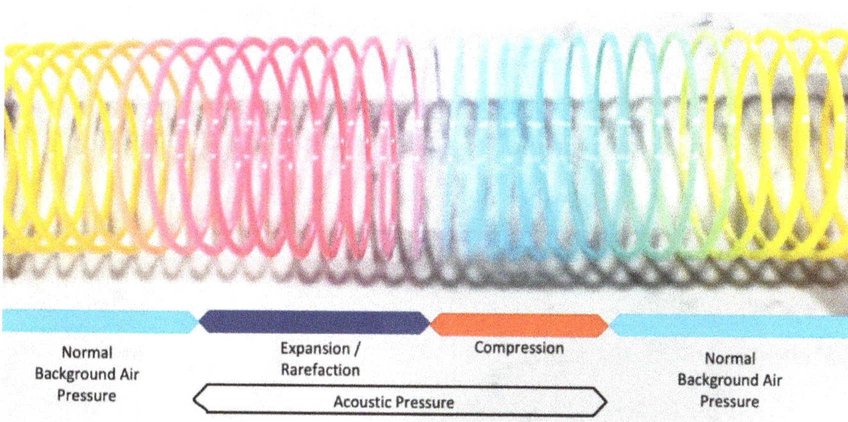

FIGURE 2 – A LONGITUDINAL WAVE; BY J. T. COOKE; 2019 © J.T.COOKE. *NOTE – THE PRESSURE WAVE PASSES THROUGH THE SLINKY IN A SIMILAR WAY, SOUND ENERGY PASSES THROUGH THE AIR.*

YouTube Channel

1.3.1 FREQUENCY

Frequency, measured in Hertz, is the number of pressure waves in a second, and the higher the frequency, the shorter the wavelength. It may be useful to think of waves on a beach and if you counted the number of waves washing up on the shore over a fixed time you might see that the smaller waves occur more frequently and larger waves less frequently. The big waves can also travel further than the smaller waves.

When trying to understand frequency, and why large waves are low frequency, and small waves are high frequency, it may help to think of the oscillation, orbit, or revolution, which created the wave. Figure 3 illustrates small to large orbits of three particles: A-blue, B-yellow, C-red, and the resultant waveform. All three particles are moving at the same speed, and the number of times the particle can complete one full rotation in one second is its frequency. The smaller orbits have a smaller circumference, and therefore A completes one full orbit more frequently. The larger orbits take longer and hence complete one full orbit less frequently. Suppose you were to shine a light onto the orbiting particles and project the shadows of the orbits onto a surface. In that case, it allows us to see the shorter and longer waveforms, as illustrated in Figure 3. Hopefully, Figure 3 gives the reader some intuition about why longer waveforms have lower frequencies, and shorter waveforms have higher frequencies. In practice, particles' vibrations or oscillations create sound pressure waves. Hopefully thinking of particles' vibrations as orbits will provide the reader with intuition as to why large waves are low frequency, and small waves are high frequency.

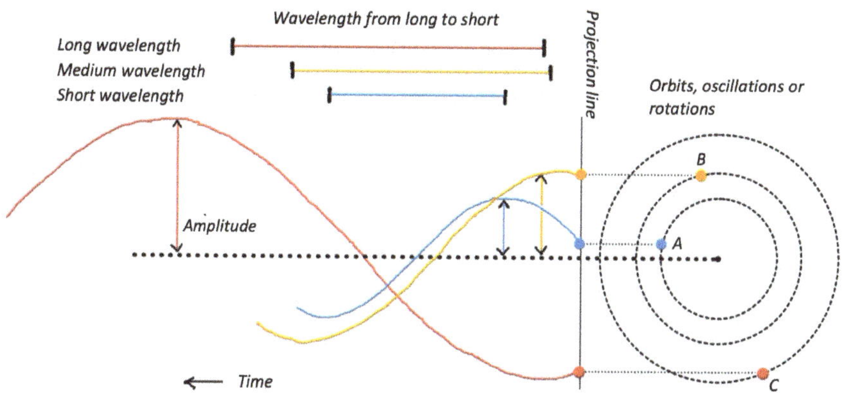

FIGURE 3 - WAVEFORMS AND FREQUENCY; BY J. T. COOKE; 2020 © J.T.COOKE. *NOTE – THE DIAGRAM ABOVE ILLUSTRATES PARTICLES A-BLUE, B-YELLOW, AND C-RED MOVE AROUND A CENTRE POINT ON DIFFERENT TRAJECTORIES. THE TRAJECTORY OF A IS SHORT, AND THEREFORE THE PARTICLE COMPLETES ONE FULL ORBIT AND ONE COMPLETE WAVELENGTH WITH GREATER FREQUENCY THAN PARTICLES B OR C. THE NUMBER OF TIMES A PARTICLE CAN COMPLETE ONE FULL ROTATION CAN BE THOUGHT OF AS ITS FREQUENCY. THE SMALLER ORBITS HAVE A SMALLER CIRCUMFERENCE, AND THEREFORE PARTICLE A COMPLETES ONE FULL ORBIT MORE FREQUENTLY. IN PRACTICE, IN SOUND WAVES, ENERGY FROM A SOUND SOURCE CAUSES AIR-PARTICLES TO VIBRATE, OR OSCILLATE CREATING THE SOUND PRESSURE WAVE WE INTERPRET AS SOUND.*

Humans perceive frequency as pitch, and a persons age can affect the range of frequencies they can perceive. The range of frequencies that perfect human hearing is sensitive to is between approximately 20Hz to 20kHz, and within this range, human hearing is more sensitive between 1kHz to 6.3kHz. To approximate for the human ear acousticians apply an A-weighting to the frequency range biased to the frequencies the human ear is most sensitive.

As illustrated in Figure 3, the higher frequencies are the orbits closer to the centre, and humans experience these as higher pitch sounds. The longer orbits are the low-frequency sounds that humans perceive as bass sounds. As noted above, the range of frequencies human hearing can typically perceive is from 20Hz, or 20 orbits per second to 20,000Hz, or 20,000 orbits per second. 20Hz is the bass sounds, and 20,000Hz is the high pitch sounds. The frequencies of familiar environmental sounds are as follows:

- Low-frequency noise is from approximately 20Hz to 200Hz, and noise from wind-farms can generate low-frequency noise.
- Dog barks can range from approximately 400Hz to 1800Hz.
- Traffic noise is approximately 760Hz to 1100Hz.
- The human voice is approximately 120Hz to 8000Hz
- Bird song is approximately 990Hz to 8000Hz

Environmental noise can be very complex and have an extensive range of frequencies, and consequently, it can be challenging to design to mitigate environmental noise. Later in the book, some rules-of-thumb for mitigating noise, such as the Quarter Wavelength Rule, will be discussed. It is essential to understand the relationship between wavelength and frequency to use the basic rules-of-thumb.

1.3.2 WAVELENGTH

The previous chapter described the relationship between wavelength and frequency to give the reader some intuition. The formula below illustrates the mathematical relationship, and it is important to be able to calculate the wavelength to understand the noise mitigation solution to choose. As we will discuss in the section about sound absorption, sound absorbers are generally more practical for absorbing high-frequency wavelength sounds. Designers need to understand the noise's wavelength and frequency to assess how they can mitigate it.

The wavelength in metres can be calculated with the following formula:

$$\text{Wavelength } (\Lambda) = \frac{Velocity\left(Speed\, of\, Sound\, in\, \frac{meter}{second}\right)}{Frequency\,(hertz)}$$

The velocity, or speed of sound in this case varies with weather and climatic conditions and an assumed speed of sound at 20° air temperature is 343 metres per second. In very humid air sound travels faster because the air is denser, it has more molecules of water that the sound energy can pass through.

For example, at a frequency of 30Hz and a velocity of 343m/s, at 20°dry air temperature the wavelength in meters would be:

$$11.43 \text{ m} = \frac{343\, m/s}{30Hz}$$

While at the same frequency of 30Hz, but with an air temperature of 30 and 80% humidity under the same atmospheric pressure the speed of sound is 351m/s, which results in a wavelength in metres of:

$$11.75 \text{ m} = \frac{351\, m/s}{30Hz}$$

So, you can see how the weather can affect the behaviour of sound. This calculation has used a standard air pressure of 101.325kPa, however atmospheric pressure can vary and further affect the behaviour of sound. When designing for something that is ever changing it is prudent to design within a range of conditions to make it more likely that a design will work with the range. It is also prudent to design in a robust way to account for variation in conditions over time. It is important to understand that acoustic calculations contain assumptions that try to approximate the measurements of an ever changing environment. The assumptions made can be incorrect and it is important to get the advice of a professional acoustician, and include calculations specific to the climate and the context of the project.

1.3.3 AMPLITUDE

The relationship between Atmospheric Pressure and Acoustic Pressure has been discussed in the previous paragraphs and Amplitude is the term for the change in acoustic pressure from the ambient atmospheric pressure levels of the context, and as there is a wide range of amplitudes when measuring pressure. If you can recall the metaphor of the atmosphere as a relatively calm sea and the swimmers splashing as the smaller acoustic pressure waves, the difference between the little waves made by the swimmers and the underlying slow rolling waves of the sea can be useful to understand Amplitude. The amplitude is the change from the underlying ambient atmospheric pressure. Human hearing has evolved to ignore the ambient atmospheric pressure, the rolling waves of the sea, and focus on the fluctuations from the atmospheric pressure, the little waves made by the splashing swimmers. Generally, humans perceive higher amplitudes as louder, and lower amplitudes as softer. If you think again of our swimmers, a large splash is a higher amplitude and gentler splashing a lower amplitude.

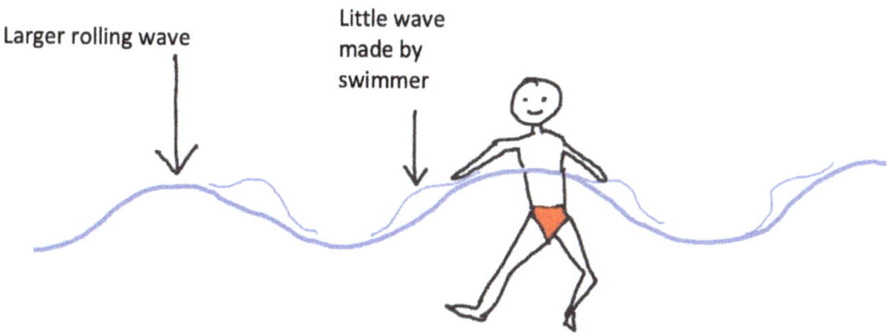

Figure 4 – Little waves on big waves; a metaphor to understand amplitude; by J.T.Cooke; 2020 © J.T.Cooke. *Note: The metaphor of the atmosphere as a relatively calm rolling sea and the swimmers splashing as the smaller acoustic pressure waves, the difference between the little waves made by the swimmers and the underlying slow rolling waves of the sea can be useful to understand the Amplitude. The difference in the height of the little waves from the underlying rolling waves similar to measuring amplitude.*

It may be useful to watch the video at the QR link below for further explanation:

YouTube Channel

1.4 ACOUSTIC METRICS & MODELS

Beware of assumptions

The weather is always changing and with it the air that sound passes through, as is the context of the sound in the Settlement & Habitation Transect. As previously mentioned, when acousticians perform calculations, they make assumptions for atmospheric pressure and a temperature range. Outside the bounds of these assumptions, the sound will behave differently to the model in the calculation. This vital part of the environmental context is sometimes overlooked when creating mathematical constructs and models as approximations of the physics of sound, and the richness of the human perception of sound. The context will affect sound levels. Hard ground, soft ground, altitude and weather conditions can affect the sound (see table 2). It may be beneficial if acoustic metrics and models consider the position in a settlement and possibly use the Settlement and Habitation Transect to categorise different metrics and models. For example, the Urban Heat Island Effect can mean that the air temperature in a city core is consistently higher than in rural areas, which affects the physics of sound. In general, rural areas will have far less deflected sound than in urban-centres. In an urban-centre a model must include for the effect of façades, as there will be greater street enclosure by buildings, and in an area with an industrial or agricultural use that makes noise once a night every 48hrs there is little use measuring the 24hr period when there is no noise.

TABLE 1 ACOUSTIC TERMINOLOGY

Atmospheric Pressure	Atmospheric pressure changes with the weather and elevation. Acoustic calculations make an assumption for atmospheric pressure. Atmospheric pressure is measured in Pascals (Pa)
Acoustic Pressure, or Sound Pressure	Acoustic pressure is the change in pressure from atmospheric pressure levels caused by sound energy. Acoustic Pressure is also measured in Pascals (Pa). Remember the metaphor of the swimmers splashing in the water.
Sound Pressure Level (SPL)	The unit of sound pressure is the decibel (dB) and is measured on a logarithmic scale.
Wavelength	As illustrated in Figure 2 the wavelength is the distance between the occurrence of rarefaction or compression in a sound pressure wave. If thinking of energy, it is the distance between where the molecules in the air have been compressed by the sound energy.
Amplitude	The amplitude is largeness of the change from the atmospheric pressure. As illustrated in Figure 2 it is a measure of the energy in the sound pressure wave that causes the change from the ambient atmospheric pressure. The metaphor of the swimmer splashing in Figure 4 may aid understanding.
dB	Decibels are the unit of measurement for Sound Pressure Level (SPL). Sound pressure level is measured on a logarithmic scale because the range between the threshold of human hearing and the limits of human hearing are so wide. A decibel is 1/10th of a bel. If you read the section on logarithmic scales, you can see that it is a magnitude scale and proportional to a base metric. The base is usually the threshold of human hearing. Decibels are a ratio, or proportion of magnitude above a base metric. Decibels are used to measure different aspects of sound and are regularly used to measure sound power and amplitude.

dB(A)	Decibel A-weighted – Humans can't hear some frequencies of sound that other animals can hear. If you have ever observed a dog barking at a high frequency whistle that a human can't hear this is because the sound frequency is outside the range that humans can hear. To focus on the audible range for humans' acousticians apply a mathematical weighting to measured sound that seeks to filter out other sound and focus on sound audible to humans. The A-weighting is common in the case of human perception of loudness, and there are other weightings, for example B, C, and D that are used to modify the measured sound for different purposes. The basis of the mathematical weighting of measured sound is arrived at by studies of people's behaviour to different frequencies of sound in laboratories and because individuals hearing differs and degrades with age it is an approximation of real human perception in our habitat.

Acoustic modelling can help predict noise levels in an environment and, under the guidance of a qualified acoustician, can guide the design and planning decisions of other professionals. In the author's experience, the architect, designer and planner must firstly rely on their common sense in understanding if an environment is noisy and design to the context. The author is not an acoustician, and acoustic metrics is the specialist field of acousticians, so the advice of an acoustician should be sought early in a project to measure and assess sound levels in an environment. When using models, a qualified acoustician should be engaged because models rely on assumptions that may not be suitable. The EU funded Project Q-Side found a significant deviation in the predicted road noise levels on the quiet façade of buildings with some acoustic models and found that using the Q-side acoustic model gave results that more closely approximated to the real world (Salomons, de Kluizenaar, Botteldooren, Wei, & Ögren, 2013). Perhaps additional metrics and assumptions more specific to the context in a settlement, and a structure to designate where they are used may be beneficial.

1.5 THE DECIBEL

How we measure sound level

The Decibel is how acousticians measure the intensity of sound. The intensity of the sound is the amount of energy it has driving it. A high-intensity sound will have more energy and a higher decibel level, and conversely, a lower intensity sound will have less energy and a lower decibel level. The Decibel is a ratio rather than a nominal, ordinal or interval measurement, and this is further explained in the next sections. As described in the following sections, the Decibel is a ratio to the base ten and is expressed on a logarithmic scale. Decibels are the unit of measurement for Sound Pressure Level and measure sound intensity from a reference level. The threshold of human hearing is usually used as the lower benchmark in the ratio.

The range of Acoustic Pressure fluctuations that the human ear can perceive is extensive and without modification would require a considerable measurement scale. It is more convenient to view the Acoustic Pressure range on a logarithmic Sound Pressure Level scale, and in this scale, the unit is the Decibel (dB). Deci is from the Latin

Decimus, meaning 10. A bel is a unit of measurement of the Sound Pressure Level, and a deci-bel is 1/10 of a bel.

Calculations with decibels can seems complex and a handy calculator is available at the following website:

noisetools.net

1.6 WHAT IS A SCALE OF MEASUREMENT?

The best form of measurement

To understand why acousticians use a logarithmic scale to measure sound it is important to understand how we measure the world around us, and the different methods of measuring. For example, if we wanted to know if Kasper is taller than Sean how can we tell? We could stand them back-to-back and see who is taller, or make two marks on a door frame, and see that Kasper is taller than Sean. Now that we know Kasper is taller than Sean we can rank or order their heights; Kasper is tallest and Sean second, and this is a basic ordinal (order) measure. How can we tell how much taller Kasper is? In order to be able to measure the relationship between Kasper's and Sean's height we need an independent scale of measurement. In the case of Kasper and Sean's height we could use a simple ruler that divides a linear amount of space into equal sections and each section is given a number, and this is a nominal (number) measure. The sections that divide up the linear scale could be anything, as long as they are consistent. The scale of measurement in a ruler to measure height could be divided into meters, centimetres and millimetres and can then be used to measure other children's heights. Once we have measured Kasper and Sean's height, we now know the ordinal measure (who's tallest, second tallest etc.) and with the nominal measure we know how much taller. If we take further measurements every year, we can observe how much each boy is growing per year from birth and this is a ratio measurement, and could compare who is growing faster Kasper or Sean? A logarithmic scale is a way of describing the ratio and can be used to measure growth rates, and often population growth rates.

1.7 WHAT IS A LOGARITHMIC SCALE?

Understanding a logarithmic scale

Most people are used to dealing with linear scales, and a logarithmic scale can seem confusing because it is not linear. Logarithmic scales are very useful when you need to measure something very long, or nominal values that range from very small to very large. Sound Pressure Level is measured in decibels and because the human ear can perceive changes in sound pressure level that range from very small to very large, acousticians use a logarithmic scale for convenience.

A logarithmic scale is a non-linear scale which means that the same increase in the input may not give you the same increase in the output. A logarithmic scale is a magnitude scale where each progression of the scale is two times, or ten times, or X times bigger than the last. To help understand its useful to begin by understanding exponents.

If you remember back to maths class and that;

$2^2 = 4$, which is the same as saying 2 x 2 = 4,

and that $2^{10} = 1024$, which is the same as 2 x 2 x 2 x 2 x 2 x 2 x 2 x 2 x 2 x 2 = 1024.

Can you see how much easier it was to write 2^{10} than have to write 2 x 2 x 2 x 2 x 2 x 2 x 2 x 2 x 2 x 2 ?

In this example the scale is a doubling of the previous increment

(i.e. 2x2 = 4, (2x2)x2 =8, (2x2x2)x2=16, (2x2x2x2)x2=32, (2x2x2x2x2)x2=64, (2x2x2x2x2x2)x 2=128, (2x2x2x2x2x2x2)x2=256, (2x2x2x2x2x2x2x2)x2=512, (2x2x2x2x2x2x2x2x2)x2=1024) and in this 2 is called the 'base'.

If you can picture a long paper measuring tape with equal increments going from 0,1,2 etc. up to 1024, now using the 'base' as 2 you could then fold up the tape so it shows 2, 4, 8, 16, 32, 64, 128, 256, 512 and 1024 you would end up with a much shorter more convenient measuring tape. The logarithmic Sound Pressure Level scale works in a similar way.

Another way to think about logarithms is if you can picture a petri dish with bacteria growing out from a core, and if the bacteria doubled each time period this would be like having a base of 2, and if they increased 10 times it would be like changing the base from 2 to 10. Exponents are just the inverse of logarithms and the exponent is like the magnitude of growth of the bacteria, and the logarithm is like the time it took the bacteria to reproduce to that magnitude. The decibel logarithmic scale uses the 'base' of 10, so each increment is an increase by a factor of 10.

1.8 ACOUSTIC MAPPING

Making the invisible visible

As mentioned previously, designers and planners must understand the acoustic context. Strategic noise mapping is very valuable to alert architects and designers to noise issues and make the invisible, visible. Several European cities publish noise maps to illustrate where designers must design for a noisy context. A strategic noise map for England & Wales has been produced by the Department for Environment, Food & Rural Affairs and can be viewed at the link below:

extrium.co.uk

Similar strategic noise maps have been produced for Switzerland and the City of Berlin, and they are a valuable resource for planners and designers. Readers can access the Swiss and Berlin noise maps at the links below:

berlin.de

map.geo.admin.ch

In Scandinavia, a free modelling software specifically for predicting road noise called Nord 2000 was created for planners and designers to predict sound levels near roads and can be freely downloaded. Easily accessible tools such as Nord 2000 and Strategic Noise mapping help inform designers' decisions during the critical early design phases to create good human habitat. Please note designers and planners should be cautious in using any software and seek a professional acoustician's advice. For places that have not yet produced strategic noise maps, see the Kuu Media website link below for further resources:

kuu-media.com

In the next chapter, I will discuss the traditional design of settlements and housing to illuminate some potential solutions.

CHAPTER 2

RELEVANT HISTORICAL PRECEDENTS

What history can teach us

The problem of noise appears to have been with humanity since we first formed urban settlements and began to live at a higher population density, but following the Industrial Revolution, machines exacerbated noise issues, particularly in urban-centres.

The reasons for humans forming settlements at higher densities is often a function of geography and proximity to a necessary natural resource, such as water. Some of humanity's first cities, Harappa and Mohenjo-Daro in the Indus Valley, have the earliest mention the author can find of a law requiring buildings facing onto main roads to respond to the context by allowing only one door and no windows facing the main roadway (Marshall, 1931). Harappa also had dwellings formed around a central courtyard with windows facing into the yard. The courtyard form allowed greater user control internally and shielded the yard from the street outside (see Figure 5). This defensive form may have been for privacy, security and to prevent any impediment of the main roadway.

The courtyard, or Peristyle typology is evident in Classical Greek and Roman Housing (see Figure 5, Figure 6, & Figure 7), traditional Tamil housing, and in 19th and early 20th century city blocks (see Figure 8, Figure 9, & Figure 10).

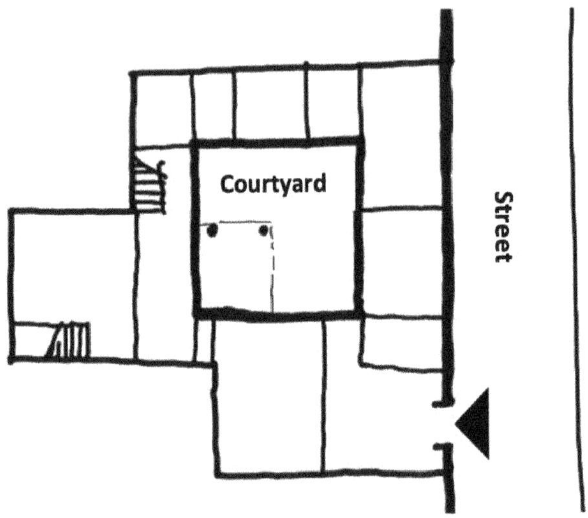

Figure 5 - Harappan courtyard house; by J.T.Cooke; May 2018 © J.T.Cooke. *Note: The rooms are arranged around a central courtyard. The building probably housed an extended family, and the yard formed a protective enclosure, providing a communal area for a family. The form of the build may also be advantageous for defence.*

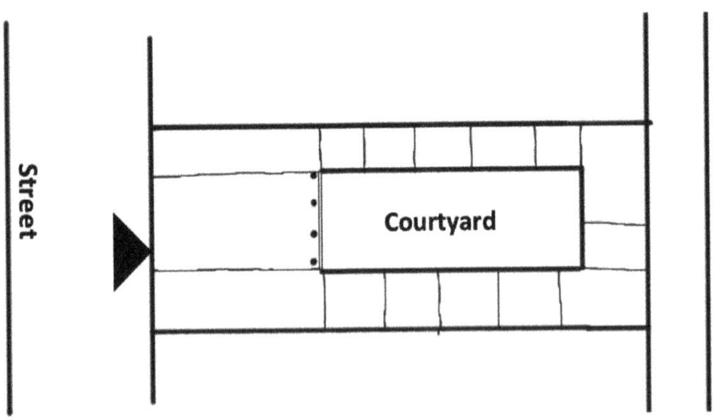

Figure 6 – Ancient Greek courtyard (Peristyle) house; by J.T.Cooke; June 2019 © J.T.Cooke. *Note: The rooms are arranged around a central courtyard. Like the Harappan house, the building formed a protective enclosure, providing a communal area for a family, and staged access from the street.*

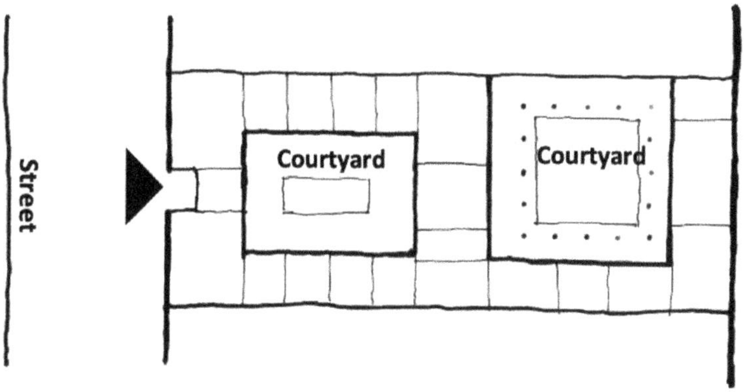

Figure 7 - Ancient roman courtyard (Peristyle) house; by J.T.Cooke; June 2019 © J.T.Cooke, based on Vitruvius description; the roman atrium house. *Note: The insular design of rooms is arranged around two courtyards and shops facing the street. Like the Harappan and Ancient Greek houses, the Roman house formed a protective enclosure, providing a communal area for a family, and staged access from the street. The building probably housed an extended family and slaves, and the front courtyard was for formal entertaining guests, and the rear yard was solely for family.*

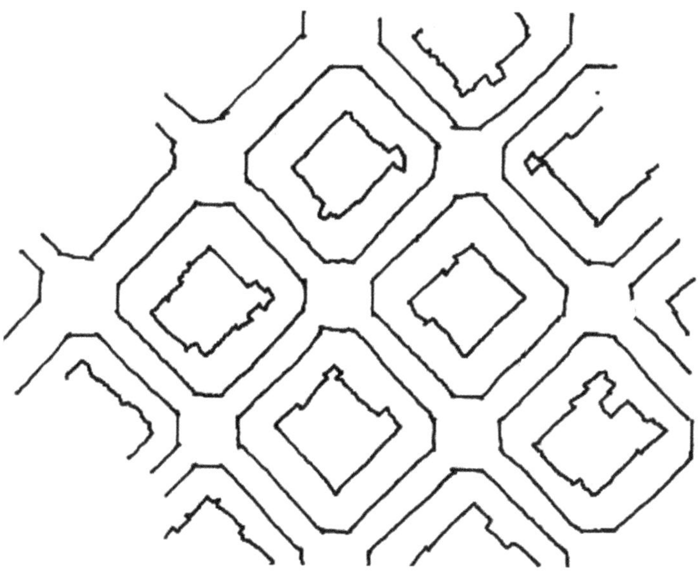

Figure 8 – Barcelona courtyard perimeter block plan; by J.T.Cooke; June 2018 © J.T.Cooke. *Note: The courtyard street block design of the new-town of Barcelona by Illdefons Cerdà created an enclosed protective space as a sanctuary from street noise and pollution. The driver of Cerdà's plan was to improve public health and give equitable access to light and ventilation to the inhabitants. There were problems with people building in the courtyards in the past, but steps have been taken to return the yards to Cerdà intended design*

FIGURE 9 – COURTYARD BLOCKS IN BERLIN, THE HOBRECHT PLAN; BY J.T.COOKE; JUNE 2018 © J.T.COOKE. *NOTE: JAMES HOBRECHT'S PLAN FOR BERLIN ALSO FOCUSSED ON PUBLIC HEALTH AND SANITATION. THE APARTMENT BLOCKS CALLED MIETSKASERNE WERE BORN OF THE APPLICATION OF SIMPLE SETBACK RULES THAT WERE INTENDED TO IMPROVE SANITATION AND ACCESS TO LIGHT AND AIR.*

In the early 20th century architects and designers changed from designing buildings that responded to the specific requirements of their context and began designing buildings that are objects independent of their context. Early 20th century architects often treated sites as a *tabula rasa*, and the author has examined many contemporary building plans using the same space planning arrangement irrespective of the context of the proposed development.

The settlement and habitation trends in the 19th and 20th centuries were the abandonment of the urban-centre and an escape to the suburbs.

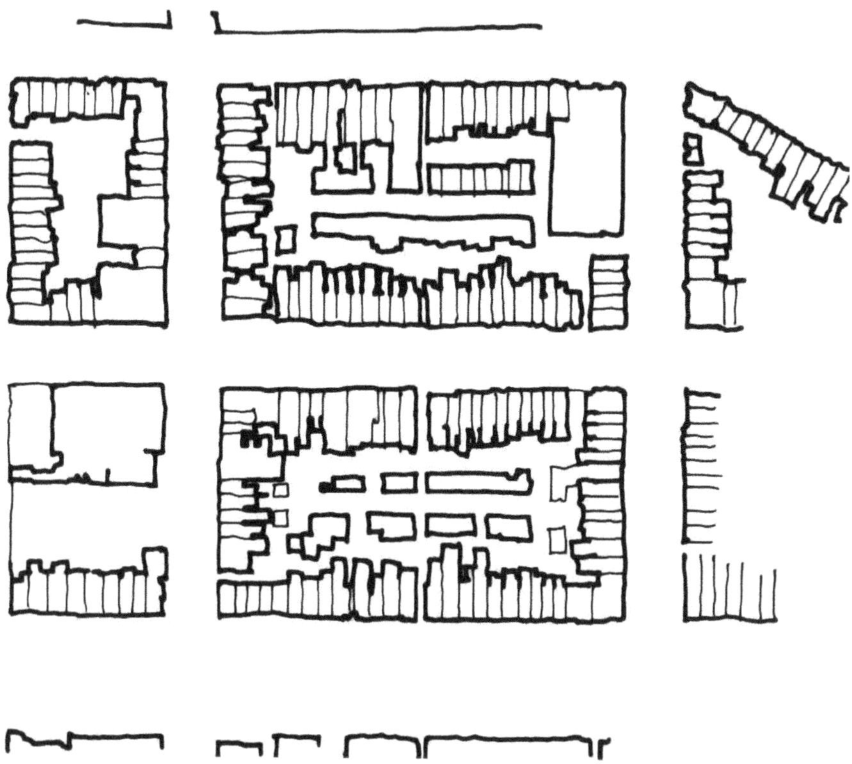

FIGURE 10 - PERIMETER BLOCKS IN WASHINGTON, MAJOR PIERRE CHARLES L'ENFANT PLAN FOR WASHINGTON; BY J.T.COOKE; JUNE 2019 © J.T.COOKE. *NOTE: HOW THE LONG INDIVIDUAL PLOTS/LOTS CAUSE AN INCREASED SURFACE AREA ON THE COURTYARD SIDE TO ACCESS LIGHT AND AIR.*

The arrival of the Industrial Revolution introduced noisy machines and polluting factories into our habitat, which exacerbated the noise and public health issues in rapidly densifying settlements. There was a widespread desire to flee the perceived toxic density of the congested, polluted and squalid urban-centre and escape the associated public health and social problems. Disease associated with living densely and pandemics, such as the Russian Flu & Spanish Flu Pandemics, also contributed to the desire to flee the city. The association of higher population density and cities with toxicity, and lower density villages with wholesomeness continues to this day. One manifestation of this desire to flee the city and seek solace in older, smaller town and village settlement forms was the garden city movement. The garden city movement was a reaction to the Industrial Revolution and the vast migration of people from the land to overcrowded cities, which were poorly prepared to house new settlers. The advent of the railways opened rural areas for the creation of commuter suburbs. The railways changed the economics of land value, increasing land values in the commuter belt and conversely, decreasing them in parts of the urban-centre. Land speculation drove railway expansion and, coupled with suburban housing this trend continued with the coming of the motor car.

Control of a right to occupy an area of land and the power over the use of that land is an inveterate precept in land law. Consequently, land-use is the primary method of control in planning law. The land-use planning solution to the public health concerns in settlements was to separate incompatible land-uses. Separation of land-uses was a robust, and well-intended solution but increased the need for people to travel to their place of work by train and then by motor car. The separation of land-uses increased the likelihood of a low-density suburban model of settlement, which relied firstly on the railways and then automobile transport and swallowed up more of our precious agricultural land. Separation of land uses cemented a reliance on motor vehicles and ironically increased the public health issues associated with noise in our cities. The increase in transport-generated noise was an unfortunate consequence of a well-meaning strategy of separating land uses.

From its inception in 1928 until its dissolution in 1959 the Congrès Internationaux d'Architecture Moderne (CIAM) significantly influenced settlement planning and formation. Settlement and habitation models, such as Hans Bernhard Reichow's new model of a Stadtlandschaft presented in his book Organische Stadtbaukunst were anti-metropolis and reliant on the motor car. The trend among these new models of a settlement sought the destruction of the dense urban-centre and can be seen as a dispersion of pockets of density into the landscape to create suburbanisation. These ideological models of settlement and habitation enabled by faster transport led us to where we are today.

The term Stadtlandschaft (urban landscape) has several meanings in the different disciplines of painting and geography and can relate to picturesque cityscapes in art, the 'natural' landscape, the changing form of landscape from urban to rural, the changing sense of enclosure, the cultural-landscape, the social/political-landscape, the character of landscapes urban to rural, and the Settlement and Habitation Transect is a tool to help classify all the manifestations of urban, rural and natural. The Settlement and Habitation Transect should be used by all involved in the creation and management of human habitat as a common reference point when communicating the context of their development. Understanding where a development sits in the Settlement and Habitation Transect can help in identifying the most appropriate acoustic design solution.

Readers can find more information in my upcoming book about Settlements, available on the kuu media website and at other book vendors.

CHAPTER 3

HUMAN SETTLEMENTS AND THE SETTLEMENT AND HABITATION TRANSECT

Characterisation of our habitat

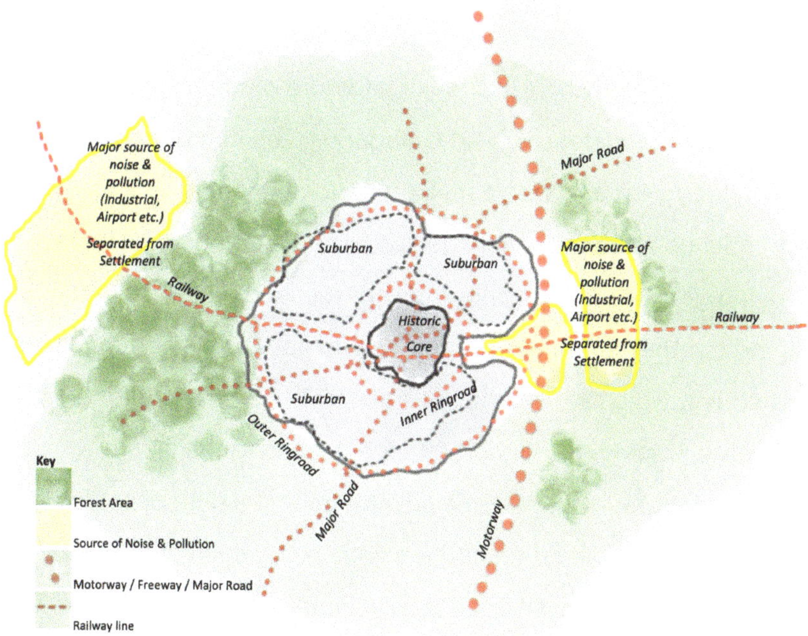

FIGURE 11 - A CARICATURE OF A SETTLEMENT FOLLOWING A CONCENTRIC GROWTH PATTERN AROUND A HISTORIC SETTLEMENT CORE; BY J. T. COOKE; JUNE 2019 © J.T.COOKE. *NOTE: IN THIS GENERAL MODEL OF A SETTLEMENTS DEVELOPMENT THE SETTLEMENT GROWS IN CONCENTRIC WAVES WITH THE OLDEST PART OF THE SETTLEMENT AT THE CENTRE, SUCCESSIVE DEVELOPMENT OCCURS OUTSIDE THE HISTORIC CORE, AND INFRASTRUCTURE INCLUDING RING-ROADS & RAIL LINES CAN BE ON THE PERIPHERY AND PENETRATE INTO THE SETTLEMENT CENTRE.*

Humans have been forming urban settlements for thousands of years, and settlements exist in a vast array of ages, sizes and geographic locations. The UN estimates that 68% of the world population will live in urban areas by 2050 [7] (United Nations, 2018), and it is critical that our settlements provide excellent habitat for humans. Conceptualising human settlements as human habitat provokes all professions to remember that we aim to create better habitat for humans. Thinking in this holistic way, the disparate professional fields of; governance, law, public health, planning, engineering, transport planning, urban design and architecture can be given a common reference point and aim.

A settlement will form, and people will choose to inhabit a place because of an underlying resource or a feature of the geography that is economically or defensively advantageous. The combination of features along with the people make a settlement and give it its *terroir.* How people come together and decide to settle combined with the underlying geography and topography of the land gives rise to the initial form of the settlement. Figure 12 illustrates some common types of settlement, which are: Maritime Coastal Settlements, River Settlements, Fortress or Hilltop Settlements, and for each, the underlying geography gave some advantage at the time of settlement and catalysed the foundation of the community, which could be, security, trade, or a convenient crossing point on a river. A settlements growth is determined by its prosperity. The form development takes is a function of the geography of the underlying land, and the modes of transport of its population. Many of our settlements gather around a source of water, often a river or a coastline, or river mouth on a coastline. London, Paris, Budapest, Berlin, and Rome all have their beginnings as river settlements. New York, Sydney, Los Angeles and San Francisco all have their origins as coastal settlements or coastal settlements at the mouth of a river, see Figure 12. The waterways were both a source of water and transport for many settlements. Still, when the transport revolutions of the railways and motorcar arrived in our settlements, they had a significant impact on the growth pattern and form of our settlements, and brought transport generated noise corridors into and around our settlements.

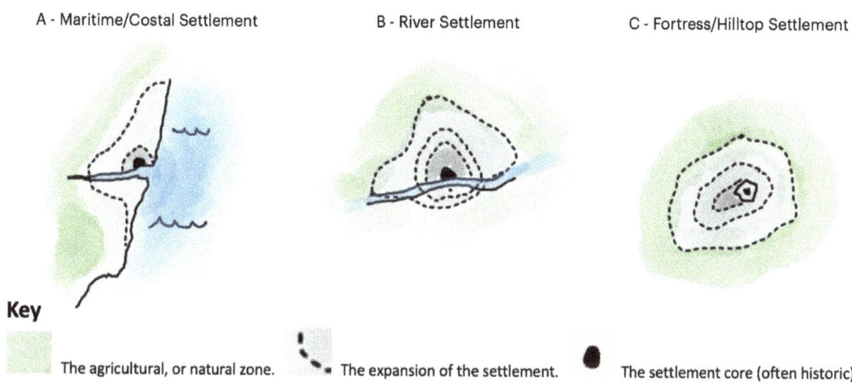

Figure 12 - Types of Settlement Form; by J. T. Cooke; June 2020 © J.T.Cooke. *Note: Settlements exist in numerous forms and some common types are: A - Maritime Costal Settlement, B - River Settlement and C - Fortress Hilltop Settlement. Maritime costal settlements usually have a seafaring origin and a port at its core, often protected by a harbour, Examples of costal settlements include: Sydney and Rio de Janeiro. River settlements usually have the origin of the settlement on one side of the river, and it may have been a crossing point on the river, Examples of river settlements include: London and Paris. Fortress or Hilltop settlements are usually inland and have a defensive structure at the settlements core, Examples of Fortress settlements include: Edinburgh and Athens. These typologies are caricatures and settlements can be combinations of all three.*

Looking at the example of the caricature of a settlement in Figure 11, a settlement forms and as time passes, the settlement expands, and the original settlement becomes the historic core. The settlement may expand outward into surrounding land, often aided by a transport network. The desire to connect the new growth of the settlement with the old growth in the centre required routes for the transport network through and around the settlement core. Transport generated noise predominantly occurs In these transport corridors, and the quality of human habitat is diminished.

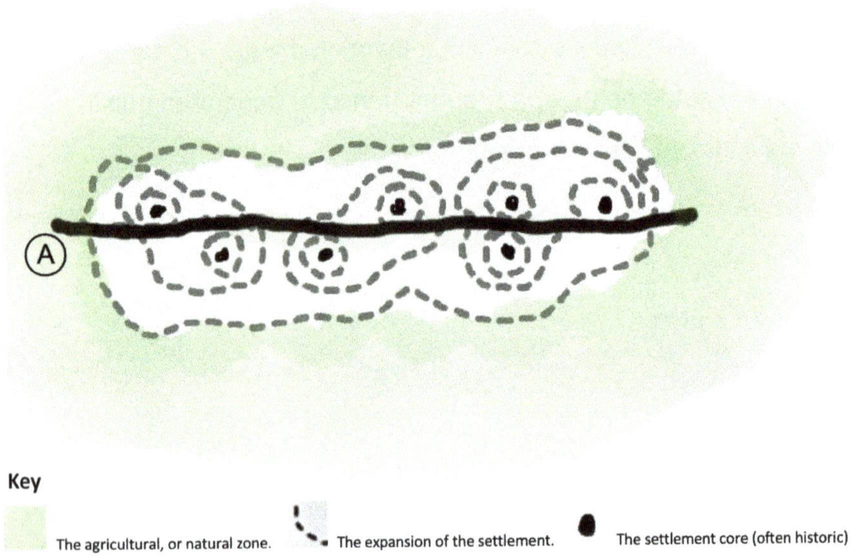

FIGURE 13 - A SETTLEMENT FORM THAT IS A PRODUCT OF THE RAIL AND AUTOMOBILE AGE, THE CONURBATION; BY J. T. COOKE; JUNE 2019 © J.T.COOKE. *NOTE: THE ONCE SEPARATE SETTLEMENTS GROW AND COMBINE ALONG THE MAIN TRANSPORT ROUTE (A)TO FORM A CONURBATION. THE POPULATION ALL LIVE ALONG THE MAIN TRANSPORT ROUTE. RIBBON DEVELOPMENT IS A SIMILAR FORM OF DEVELOPMENT BUT AT A SMALLER SCALE, AND CAN OCCUR AS A RIBBON OF PROPERTIES ALONG A ROAD.*

In many European settlements, the historic centre has narrow streets, which reflected the predominant transport of the age and were for walking and horse-drawn carts. The development of railways allowed settlements to grow and created transport corridors to the centre of the settlement, often ending in a central train terminus station near the settlement core. The later arrival of cars and motorways enabled further growth of the settlement, but once again, the desire to connect too, and through, the settlement core led to large ring roads around the centre and sometimes large roads through the settlement core, and conurbation beyond the settlement core along the transport routes, see Figure 13. The transport corridors brought noise into the settlement core along the transport routes and created a river of noise-compromised human habitat.

In Europe, many city settlements have major ring roads encircling the settlement and may have significant roads or rail lines reaching deep into the core of the settlement bringing noise with them. Many roads were built as a reaction to traffic resulting from increased motor-vehicle use, trying to access the settlement core that inhabitants had built for horse & cart, and walking. Ring roads, arterial roads, and motorways were seen as a solution to motor-vehicle congestion. Look at the following examples of cities with ring roads and rail lines reaching deep into the core of the settlement.

3.0.1 LONDON EXAMPLE

London is a river settlement located on the River Thames, and the initial settlement was on the north of the river. In the 19th Century, the city expanded rapidly, and the advent of railways brought rail lines and the associated noise deep into the city. The rail lines ended in the large terminus stations of; Waterloo, Euston, St Pancras, Kings Cross, Liverpool St, London Bridge, and Paddington Stations clustered close to the settlement centre.

The subsequent rise in motorcar use brought an increase in roads and traffic. Four ringways or ring roads were proposed in the early 20th

Century to ease congestion, but the authorities completed only three. There were significant public protests about the noise, air pollution and disruption the ring roads brought. The current ring roads are the Inner Ring Road along the A501 around the old city core, the North & South Circular Roads along the A406, A1020, & A205, and the M25 forms the most recent ring road. Other motorways and arterial roads travel deep into the settlement bringing vehicle noise with them. It may be helpful to look at the free map at the link below:

openstreetmap.org

3.0.2 PARIS EXAMPLE

Paris is also a river settlement and began on an island, the Île de la Cité in the River Seine. The advent of the railways in Paris came at a similar time to that of London, and the railways brought rail lines deep into the historic settlement of Paris. The rail lines ended in the large terminus stations of; Gare du Nord, Gare de l'Est, Gare d'Austerlitz, Gare de Bercy, Gare de Lyon, Gare de Montparnasse, and Gare Saint-Lazare, clustered close to the settlement centre.

Like London, the rise in motorcar use led to an increase in roads, ring roads, and arterial roads that reached deep into the settlement core. The current ring roads are the inner ring road, the Boulevard Périphérique (Périph), the A86 or Paris super-périphérique, the Francilienne a partially complete ring road, and the partially complete Grand Contournement de Paris. Other arterial roads and motorways go deep into the settlement bringing vehicle noise with them. It may be helpful to look at the free map at the link below:

openstreetmap.org

3.0.3 MUNICH EXAMPLE

Munich is a river settlement on the River Isar. Like London and Paris, but at a lesser scale, the expansion of the railways brought rail lines and associated noise into the settlement terminating in München Hauptbahnhof (Munich Central Station), München Ost, München-Pasing, and München Nord.

Similar to London, and Paris, the increase in motorcar use, brought an expansion of road building. The current ring roads are the Altstadtring, which is the innermost ring road, the Bundesstraße 2R or Mittlerer Ring (Middle Ring), and the partially complete Bundesautobahn 99 outer ring road. Other arterial roads penetrate deep into the settlement bringing vehicle noise with them. It may be helpful to look at the free map at the link below:

openstreetmap.org

Hopefully, you could identify the areas with ring roads, major roads and rail passing through the settlements. Looking at these examples, it is clear that the same transport technology and transport solutions create the same noise corridor issues in our settlements.

Science uses the cross-section to see what is hidden and simplify complexity. The father of science, Aristotle, used vivisections regularly to understand anatomy. In geography and ecology, a transect is comparable to a section through land and habitat and simplifies complexity. Alexander von Humboldt was the first to use a transect to clarify and understand the complexity of the geography of the south of South America in the 18th Century in his work Tableau Physique. Andrés Duany and Elizabeth Plater-Zyberk promote the use of transects in urban design and urban planning through their practice DPZ and the Centre for Applied Transect Studies (Duany Plater-Zyberk & Company, 2019). The transect is helpful because it simplifies the complex and is easy enough to understand to be used as a basis of discussion for a vast audience.

A Settlement and Habitation Transect may be a way to simplify, categorise and understand a settlement and can be used to find common patterns in our habitat. The author has produced a general Settlement and Habitation Transect in Table 2. The transect illustrates how settlements generally tend to be denser and cityscape-dominated in urban centres and less dense and landscape-dominated in rural areas at the settlement periphery. The reasons urban centres are denser than peripheral rural areas are many and include proximity to desirous resources, which today may be jobs and services. Proximity to what is desirous drives land values and density higher, and what is desirous changes over time.

The different lands, cultures, history, latitudes, and geography of the earth have produced many settlement forms. As a word of warning, this richness should not be over-simplified to fit a settlement model. Still, while keeping the reality of complexity in mind, simplification can be helpful to aid understanding. A general settlement and habitation transect should not be imposed on a settlement and is a tool to analyse and understand the uniqueness of a settlement.

As illustrated in the examples, the transport corridors inevitably deliver inhabitants to what is desirous and create noise compromised territory. The transport corridors could be classified as a particular character in the Settlement & Habitation Transect. Development in transport corridors is often affected by transport-generated noise. City planners could give these types of territory in our settlements a designation, such as conurbation, which reflects their character and guides the forms of development and noise solutions that may be more suited to the context.

Many settlements have areas that are blighted by proximity to noise from roads, railways, or industry. When all the best human habitat for housing has been used, strategic planners seek to grow settlements through consolidation and often look to land previously rejected as sub-standard human habitat, such as old industrial areas. In countries with market economies, where the land & property market is subject to speculation, land-value economics mean that landowners and developers will only want to re-develop land to a land use with the highest financial yield. Landowners and developers will usually seek to buy cheaper land and re-zone it to a land-use that provides the highest monetary return.

In the post-industrial nations, the decline in the need for land for industrial, low-grade commercial, and small retail uses, together with an increasing demand for housing, has driven pressure to build housing on old industrial, low-grade commercial and small retail sites. Land-value economics has meant that re-development of cheaper industrial, low-grade commercial and small retail land to housing has been occurring in many post-industrial settlements.

Small retail and commercial land-uses are often found along noisy old main roads. The desire to consolidate existing settlements has led to increased pressure to redevelop lower financial yielding small retail and commercial land-uses into housing to gain a higher financial yield. In these noise-prone areas, if the new dwellings, often high-density housing, are not designed correctly, it can create a very poor human habitat. Similarly, ring-roads around settlements often produced a halo of noise-affected land that provides humans with poor living habitat, see Figure 84. Some good examples of designs to improve human habitat near Munich's ring road are discussed later in the book.

As mentioned, another common trend in post-industrial settlements has been the reduction in industrial land-uses. Combined with a desire to consolidate existing settlements, and increasing demand for housing has led to increased pressure to re-use old industrial land to provide housing. The previous industrial uses often contaminated Post

-industrial, or Brownfield sites, which are often redeveloped as high-density developments to give enough financial return to pay the decontamination costs. The old industrial, low-grade commercial and small retail sites are usually located in areas of our settlements that are compromised by noise and need a design response appropriate to the context.

In the cities of the post-industrial world, there are many examples of new housing built on old industrial land and adjacent to industrial land that is still in use. Locating new housing near industrial uses can cause noise problems for the new residents if the building is incorrectly designed. The following are examples of areas where new housing is located close to, or in, industrial & former industrial areas. It may be helpful to look at the free map at the links below:

3.0.4. GREENWICH PENINSULA LONDON EXAMPLE

In this example, new housing has been located near the Greenwich Peninsula Ecology Park and to the west of an industrial area.

openstreetmap.org

3.0.5. HAFENCITY HAMBURG EXAMPLE

In this example, the new housing district of HafenCity is located to the north and west of industrial areas and the operational commercial port. HafenCity is situated close to central Hamburg, and a railway line is located to the east of the area. The noise from the port and industrial areas presented a challenge to the city in the development of HafenCity, and I will discuss some innovative solutions later in the book.

openstreetmap.org

3.0.6. GREEN SQUARE SYDNEY EXAMPLE

The Green Square Urban Renewal Area is a new precinct consisting primarily of high-density housing. The planned core of the area is located around Green Square Station. The area is an old industrial zone in the south of Alexandria, Sydney. To the east and south of Green Square Station, the old housing stock, which was located within the industrial zone, will now be incorporated into new high-density residential areas. Busy roads bisect the new housing areas and present a challenge to designers and planners.

openstreetmap.org

When planning and designing new settlements, or areas within existing settlements, some transport engineers may look at context through a different lens from architects, urban planners, or a concerned neighbour. Lack of shared understanding of context can lead to misunderstanding why a noise solution may be appropriate in one context but not in another. Lack of a framework for context has led to guidelines for acoustic design that are often one-size-fits-all solutions primarily developed for suburban contexts. In urban-centres separation solutions may not be practical, nor appropriate and it may be more beneficial to remove the source of the noise for high levels of noise (e.g. motorway, aircraft noise) and reduce levels of noise (e.g. human activity noise, music, school playgrounds) with the use of barrier and absorption solutions (see Table 2).

No individual can accomplish the ongoing successful growth of a settlement, and it involves the community, landowners, new settlers, politicians and all disciplines involved in the creation, management and maintenance of settlements. A competent team, the community's goodwill, and a common reference point are required to create good settlements and excellent human habitat. Unfortunately, those involved in creating and managing settlements and the community spend too much time arguing about context and the appropriateness of proposed developments to context.

It may help general understanding to designate the character of areas using the Settlement and Habitation Transect, assigning some areas as conurbation etc. The character areas in the Settlement and Habitation

Transect could be a standard tool to frame interdisciplinary and public discussions.

		Urban to Rural	City / Townscape to a Landscape Dominated	Standard Noise Mitigation Solutions	Building Stock Age	Acoustic Attenuation Solution Consists Primarily of:	Acoustic Context
Settlement Ancillary Area		*Ancillary; satellite to settlement*	*Ancillary; satellite to settlement*		*Varies – often an airport, or industrial area. Can become a site of urban renewal and densification.*	*Separation*	*Varies*
Rural & Agricultural Area	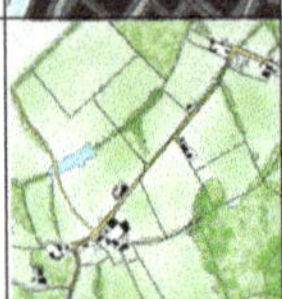	*Rural*	*Landscape Dominated*		*Old stock in the historic settlement often with large suburban proposals nearby*		*Generally, more soft ground attenuation, and cooler than the urban core. More vegetation is mitigating noise.*
Sub-Urban Area		*Sub-Urban*			*Varies - newer stock with infill development of existing suburbs*	*Barrier*	
Conurbation & Ribbon Development Area	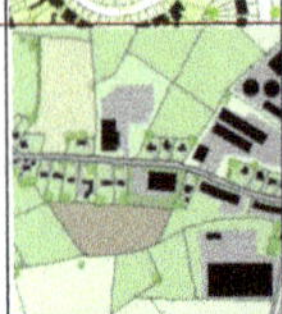			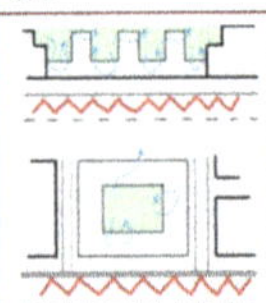	*Mixed. In rural ribbon development usually new stock, in the renewal of old industrial area usually newer apartment stock, conversions.*		*In existing settlement peripheral areas (old industrial areas etc.) Acoustically harder ground. In more rural areas is similar to rural & agricultural areas.*
Neighbourhood Centre Area				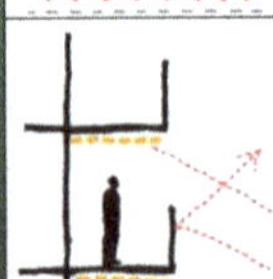	*Generally older stock with mainly infill development*	*Absorption, Diffusion and Barrier*	*Acoustically hard ground. Usually warmer because of Urban Heat Island Effect, potentially greater temperature inversion at night bending noise down to receivers. Buildings along streets create canyons reflecting and reverberating sound more than in rural areas.*
Settlement Centre Area	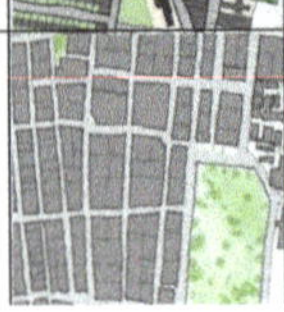	*Urban*	*City/ Townscape Dominated*	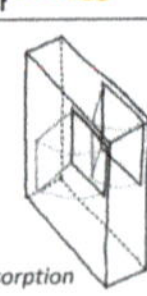	*Oldest stock in a historical core with mainly in-fill development. Sometimes a CBD with newer stock.*		

TABLE 2- THE SETTLEMENT AND HABITATION TRANSECT FROM URBAN TO RURAL WITH CHANGING CHARACTER AND NOISE MITIGATION SOLUTIONS; BY J.T.COOKE; JUNE 2018 © J.T.COOKE.

If you have found this section engaging and would like to read more on this topic, you may find my upcoming book about Settlements interesting. Further links and information is available at www.kuu-media.com

3.1 NOISE AND HUMAN HABITAT

Noise and context

When creating settlements, we must aim to create an excellent habitat for humans. Sound is a part of life in human settlements and particularly in urban-centres. The decision, when a sound is loud enough or unpleasant enough that it causes a person disturbance, is subjective. Because of this, it is critical that the control over the noise that individuals are subjected too in their habitat, is given to occupants.

The context within a settlement varies across the Settlement and Habitation Transect, and research suggests that people's expectations of noisiness also change. Bruce and Davies (2014) found that the perception of noise in soundscapes was driven by prior experience and expectation, more than the perception of individual noise sources. The expectation of noise does not mean we accept low standards and high noise levels in urban-centres. Context and human perception are vital to understanding how people are affected by noise; for example, a dog barking in a quiet suburban area can annoy as much as loud noise where consistent background noise levels are higher in an urban-centre. Fastl and Zwicker (2007) and Basner and McGuire (2018) found that sudden changes in noise level are more likely to wake people than a consistent low-level white noise. The National Academy for Engineering noted,

> Context, expectations, and people's experiences and circumstances all affect their responses. Hence, levels of community response (e.g., annoyance) may vary from community to community, just as individual responses vary from person to person, even if noise levels do not change. However, one consistent finding has been that changes in noise exposure do affect individual and community responses and that increases in man-made noise usually have a negative impact. [10]

(National Academy of Engineering of the National Academies, 2010, p. 19) © National Academy of Sciences.

As discussed in *Chapter 1 The Phenomena of Sound*, the physics of sound and the human perception of sound are complex, and because of this, defining a metric to assess noise is difficult, and no single metric can work in all cases. Human settlements can be noisy and can range in type and intensity of noise throughout a settlement, and this is why design solutions must give choice and control over subjugation to noise, to the occupant.

3.2 SOURCES OF NOISE IN SETTLEMENTS

Living with noise

Noise in human settlements emanates from many different activities and is a part of the life of a village, town or city. Various sources of noise such as road noise, rail noise, or aircraft noise can be thought of as having different *sound profiles* and understanding this can help choose the most appropriate solution. Sounds are a mixture of sound waves, and some have more low-frequency bass components, and others have more high pitch, high-frequency components. The *sound profile* is the particular combination of pressure waves and frequencies that our brain then identifies as a train, or car, or aircraft etc. Noise mitigation solutions may be better suited to low, or high-frequency sound, and you may need combinations of solutions to deal with some noise sources. In human habitat sources of noise fall into two main categories, which are:

Internal noise sources from activities within buildings; neighbours, babies crying, loud music, and are mainly transferred through the building fabric, and;

External noise sources from activities outside the building envelope, which are mainly transferred through the air via ventilation openings in the building envelope. The external noise sources can be further sub-categorised into the main external noise generating activities of a settlement, namely:

> ***Machines*** (mainly transport): road vehicle noise, railway noise, aircraft noise, etc.
>
> ***Industry***: concrete batching plants, factories, etc.
>
> ***People***: including leisure activities, school playgrounds, street life, etc.

The land-use planning system has historically done a reasonable job of separating industrial areas and the associated noise from noise-sensitive areas. However, transport-generated noise is more difficult because it connects and passes through, or over settlements. As discussed in Chapter 3 and illustrated in Figure 11, our settlements grow and change over time, and new forms of transport enable a settlement to spread across the land. The new methods of transport sought access to the urban-centre of settlements, and in particular, the railways and automobiles transported noise into our settlements along transport corridors. In the dense urban-centre of settlements, where separation from road and rail noise is not always practical, there are two options:

> **Remove the source of the noise** by removing machines, automobiles, and trains from the urban-centre, and potentially locating the trains under the urban-centre as a metro.

> **Control the noise through the planning system**, and design to give people control over the noise they are subjected to, until the removal of transport noise in the future.

3.3 NOISE AND PUBLIC HEALTH

The case for action

A large body of research and evidence exists that highlights the health impacts of living in a noisy environment. Legislation and public policy have sought to control the detrimental effects of noise by controlling the noise-level and the duration of exposure to noise. Recently, the World Health Organization (WHO) produced *Environmental Noise Guidelines for the European Region* (World Health Organization, 2018) which is a rigorous, public-health-based document to guide policy-making and improve public health.

Another WHO report, the Burden of disease from environmental noise quantifies the negative effects of long-term exposure to road traffic noise levels. In the EU States, the WHO report estimates the 'disability-adjusted life years' (DALYs) from noise-related issues, such as:

ischemic heart diseases (61,000 years)

cognitive impairment of children (45,000 years)

sleep disturbance (903,000 years)

tinnitus (22,000 years)

annoyance (654,000 years) [2]

(World Health Organization Burden of Disease from Environmental Noise; Quantification of healthy life years lost in Europe, 2011, p.101) © WHO Copyright 2011; Reprinted with the permission of the World Health Organization.

Van Renterghem and Botteldooren (2012) estimate this to be a loss of about one million healthy life years, every year in Western Europe. Basner and McGuire (2018, p. 38) noted, 'There is biologic plausibility that chronic night-time exposure to relevant levels of noise can contribute to negative health consequences like a cardiovascular disease'[1]. The UK Department for Environment, Food and Rural Affairs (2018) estimates the social cost of road noise in the UK at between £7 billion and £10 billion per annum, and also estimates the financial cost of noise to be between £60 billion and £100 billion per annum with sleep disturbance the single highest health impact. Given the cost to the taxpayer and relatively inexpensive cost of designing for noise control it would be prudent to insist on better acoustic design.

Research points to the benefits of having a quiet façade, which could be provided by a perimeter block urban form (see Figures 94-96). A study by de Kluizenaar et al. (2013) supported previous research by Öhrström et al. (2006) and, corroborated by research by Van Renthergem and Botteldooren (2012), noted that residents exposed to traffic noise benefited from having a quiet side in their home where they could escape noise.

Similarly, an EU funded project called 'Project QSIDE' investigated how European cities can reduce the harmful effects of traffic noise by providing quiet facades and quiet urban areas as a refuge for residents. It concluded that 'quiet places' (private quiet facades, and private quiet urban areas) are beneficial for the inhabitants' health, and we will explore designing buildings with quiet sides later in the book.

Another EU funded project, the SONORUS (2016) project, investigated the role of perimeter block urban forms in offering a quiet side, and it recommended designing a courtyard to provide a high-amenity quiet refuge. As discussed in section *1.4 Acoustic Metrics & Models*, the definition of what constitutes a quiet facade varies and some studies have variously measured A-weighted sounds levels over; 8hr, 24hr and 48hr periods and slightly differing decibel levels, and some acoustic models include reflected sound while others reviewed do not. De Kluizenaar et al. (2013) indicated that lower noise levels on the least exposed façade might provide a more significant health benefit for inhabitants. Conversely, higher noise levels on the least exposed façade may increase health problems, indicating that the tranquil quality and quietness of the courtyard's 'quiet side' may be more beneficial for the health of inhabitants.

As illustrated in the section on historical precedents, protective courtyard forms are not new to architecture and planning. From the Ancient Greeks to Ildefons Cerdà's courtyard city blocks in Barcelona, there are many examples of courtyard forms that create a sanctuary for a family or community, see Figure 6 & Figure 8. Architects and planners widely used courtyard forms for housing in the past, but arguably less so since the early modern movements preoccupation with cheap housing towers and the creation of buildings as objects in the landscape, devoid of any relationship to their context.

It is in the interest of public health that designs respond well to the context in a settlement, and acoustic context, and it is essential to understand the potential solutions available. I will discuss the practicalities of developing perimeter block urban forms and courtyard buildings in section *6.3 Site planning,* and in the next chapters, the author will discuss designing for noise control and illustrate some innovative examples.

3.4 NATURAL VENTILATION

A brief explanation

FIGURE 14 – A LANEWAY IN SINGAPORE; BY J.T. COOKE; SEPT 2018 © J.T.COOKE. *NOTE: THE REAR LANE FAÇADE OF THE BUILDINGS ARE ALMOST ENTIRELY COVERED IN AIR-CONDITIONING CONDENSER UNITS.*

People spend much of their lives inside, and the Covid-19 Pandemic focused attention on the air quality in our buildings. The Pandemic highlighted the need for a plentiful supply of clean, fresh air, and early research suggests the benefits of natural ventilation.

Basic Principles

- The natural forces that make air move in natural ventilation include the following;

  - **Pressure difference** - A pressure difference from one side of the building to another. The greater the difference, the greater the movement of air. The pressure difference can be created by the following: wind, a difference in temperature from one side of the building to the other, difference in internal & external pressure. Excessive ventilation rates can cause problems like blowing papers off desks, and unevenness of occupant comfort across a room.

  - **Air buoyancy** - The buoyancy created by heat from occupants and heat-generating appliances warms the air, which rises drawing more air into the vacuum created by the rising air. Heat from the sun can also create air buoyancy.

- **Single or Dual-Sided** - Single-sided with an opening on one side of the building will ventilate a limited depth of a room. Dual sided with openings on two opposite sides provides natural cross-ventilation, and can ventilate greater building depths given the right weather conditions.

- **For respiration** – sufficient changes of air while not losing heat or coolness.

- **For cooling** – sufficient flow of air for evaporation of sweat from a person's skin (evaporative cooling) and cool the building fabric. Night cooling can be an effective strategy.

- **Optimising opening size** - Designers should tailor the opening size to the ventilation rate required. The weather conditions will determine the air's velocity, so openings need to be adaptable by the occupant to tailor the opening size as weather conditions change.

Natural ventilation is sometimes referred to as passive ventilation, and is the use of passive natural forces to ventilate a building rather than using electric fans, or active energy consuming mechanical ventilation systems. There are many building designs and devices that seek to harness natural forces for natural ventilation, which include; windcatcher chimneys, chimneys, Trombe wall solar air pumps, and providing openings on opposite sides of a building.

There is a spectrum of ventilation solutions from totally passive natural ventilation, to hybrid ventilation where some mechanical ventilation and energy is used, to complete HVAC systems. Extract fans in bathrooms and kitchens are one such driver of air movement in a hybrid system, and the extract fans create a *negative pressure* internally, which draws more air in from outside through openings. The air-tightness of buildings is key to controlling air infiltration, and the locations where air can enter a building.

Natural ventilation does not require electricity to work and has low energy and maintenance cost, and is sustainable. Prior to the advent of mechanical ventilation and HVAC systems all buildings were naturally ventilated. Mechanical ventilation and HVAC systems require electricity to work and can have high energy consumption and high costs. Cheap energy in the 1960's drove the expansion of the use of mechanical ventilation and HVAC systems, and as energy costs rise the systems become more costly and less attractive to building owners, and commercial and residential tenants. HVAC, mechanical ventilation and electric lighting enabled building floor-plates to deepen, with a small ratio of building surface area to floor area and were thought to be space efficient. Naturally ventilated buildings usually have a higher ratio of building surface-area to floor area because the rooms are ventilated from the external surface of the building. Later in the book, in section *6.3 Site planning,* a return to increased surface area of a building on the quiet side is illustrated in Figure 94.

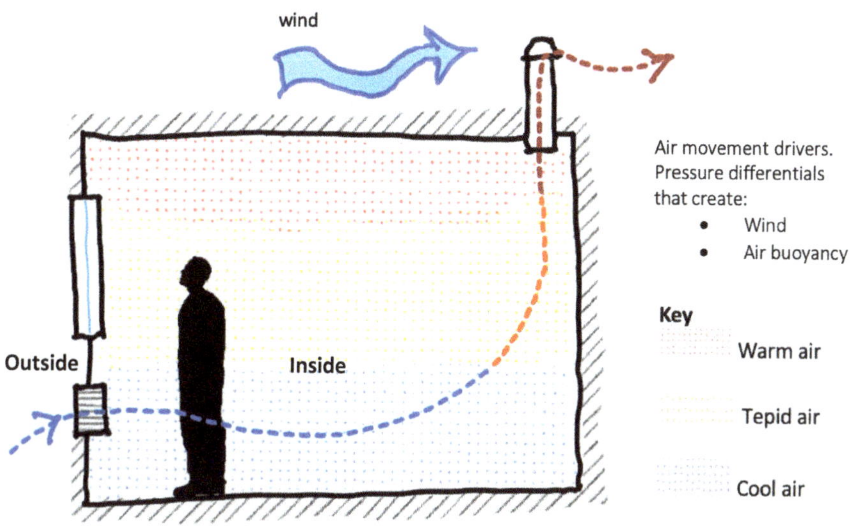

FIGURE 15 - THE DRIVERS OF NATURAL VENTILATION; BY J.T. COOKE; NOV 2020 © J.T.COOKE.
NOTE: THE IMPORTANCE OF THE LOCATION OF THE OPENINGS DEPENDING ON THEIR FUNCTION AS, INTAKE OR EXHAUST AIR.

Both HVAC systems and natural ventilation are used in buildings to supply clean fresh air for respiration (people breathing) and for cooling the occupants, in addition, HVAC systems also heat & cool the air of buildings. A natural ventilation strategy for a building should be carefully planned and designed, and not simply an opening in the external façade. The poor design of natural ventilation can lead to occupants retrofitting mechanical ventilation, or HVAC systems to make their habitat more comfortable. Noise from outside a building generally enters the building through openings in the external façade for natural ventilation, usually windows, balcony doors and rooflights. You can read more in my book about natural ventilation available at www.kuu-media.com.

3.5 SUSTAINABILITY AND ACOUSTIC PRIVACY

Noise and Overheating

It is self-evident that providing cooling for buildings through natural means that do not require energy is the best option, the most sustainable and the lowest cost. To mitigate overheating in their homes, occupants usually attempt to increase ventilation rates using purge-ventilation, by fully opening their windows. In locations with high external noise levels and poorly designed buildings, occupants may have to choose between open windows and high noise levels, or closed windows and overheating. It is not good design to assume that if a dwelling has windows, that the open windows will provide natural ventilation for cooling. It is bad design to make people choose between an overheating dwelling, or acoustic privacy when both are possible in tandem with a good design.

It is particularly important to consider overheating, natural ventilation and acoustic privacy within urban-centres, and high-density housing. In urban-centres, the closer proximity of dwellings invariably provides a more significant opportunity for a noise disturbance, and in particular where natural ventilation and night-time cooling is the inhabitants' primary method of temperature control.

There are numerous natural ventilation cooling strategies, and the most appropriate will depend on the local climate. A standard cooling approach in climate zones other than tropical zones is 'night-cooling'. Night-time temperatures can be significantly lower than the daytime temperatures, and the cooler night-time air cools the occupants and the building fabric, leading to a cooler internal building temperature during the day. If proposing a night-time cooling strategy in dwellings, an architect must consider the increased likelihood of sleep disturbance because of the lack of an acoustic barrier. Because night-time cooling relies on the opening for natural ventilation being open (i.e., acoustically compromised) at the time when inhabitants require the highest levels of acoustic privacy (i.e., while they sleep), this presents designers with a problem, but there are solutions, which will be shown later in the book.

When proposing natural ventilation as a cooling strategy, a designer should carefully consider the design of the building openings that will provide the ventilation and tailor them to suit the context, and this is discussed in section *5.6 Locating Ventilation Openings and Windows*. In warmer climates, the required ventilation rate may be high and subsequently, the ventilation opening more substantial, and these openings need to be well designed. One tactic is to have separate building envelope openings to provide light and natural ventilation, which allows the designer to tailor the design of the two different openings to the specific requirements for light or natural ventilation. For example, using a sealed double or triple-glazed window on a facade that receives sunlight and noise, but the opening for ventilation could be located away from the source of noise and designed to remove noise while allowing natural ventilation. Figure 35 to Figure 37, illustrate some examples of this tactic.

Architects, designers and acousticians should not default to specifying air-conditioning, nor mechanical ventilation, and therefore pass the cost for poor design on to the inhabitant in higher energy bills; it is bad for the consumer and terrible for the environment, and inherently unsustainable and unnecessary.

3.6 CONTROLLING NOISE IN SETTLEMENTS

Noise and Planning

Land-use planning, and zoning restrictions have sought to separate industrial land-uses to keep the noise and pollution created away from people's homes. As previously discussed, land-use planning has also tried to keep transport-generated noise away from people's homes. However, control of transport generated noise by separation alone is problematic. Separation by land-use is a robust solution because it mitigates a wide range of problems, including; noise, pollution, outlook, and views, but it has inadvertently exacerbated the problem of low-density sprawling settlements that rely on automobiles to transport the inhabitants.

The increasing population density of our cities mean that we cannot rely on the separation of land-use alone to mitigate noise problems. In recent years, planning has sought to reduce dependence on cars and improve sustainability by the consolidation of our settlements. The land used for consolidation is usually land that was rejected for housing development in the past because it may not have been suitable human habitat, or more suitable for another use. Settlement consolidation sites usually present design challenges to be made suitable human habitat, and noise is often one challenge. The decline of industry in Europe, the USA and Australia has created pressure to change land once saved for industry into land for housing. In many settlements, the prime housing land is already built upon, and industrial areas once thought unsuitable for dwellings is being developed for housing, as illustrated in the examples at the beginning of the chapter. These sites are often peripheral to the urban-centre in general-urban areas, and can be close to rail or roads because this was necessary for the industry. The proximity to road and rail often brings associated noise problems when these brownfield sites are being redeveloped into housing.

Similarly, the case for sustainable Transit-Oriented Development (TOD) is premised on compact, dense settlements. It relies on locating people closer to public transport nodes such as train stations, but requires good design to create suitable human habitat. One common design solution in over train-station, TOD is to separate residential uses from the noise of the railway by including floors of less noise-sensitive uses between them.

In densely populated urban-centres, we must look to design solutions, specific planning controls, and removing transport-generated noise from urban-centres. The trend toward urbanisation means an increasing number of the world population will live in dense settlements, and we need more precise planning controls to deal with the complexity and density of urban-centres. In the author's experience, if we had space-use controls rather than land-use controls, it would be possible to regulate vertical levels of space and different areas within buildings. A more precise level of space-use control is more appropriate in dense urban-centres, rather than land-use, which has its origins in agricultural land-use and a rural context. Land law is rooted in our agrarian past, and the substitution of 'land' for the word 'space' in land law may allow a customisable level of control for the location in the settlement and habitation transect. As discussed previously, TOD developments often separate residential uses from the noise of the railway by including floors of less noise-sensitive uses between them. If land-law controlled 'space' rather than just 'land' more precise separation strategies could be allowed for and managed, giving more certainty that development will create good human habitat.

CHAPTER 4

DESIGNING FOR NOISE CONTROL

Designing with the invisible in mind

Architects & Designers often overlook noise because of the primacy of the visual sense. We market buildings on their image and seldom mention the acoustic quality, and inhabitants only realise acoustic problems after occupancy of the building. Architects and Designers must firstly use their sense of hearing when they visit a site and respond to the context for all the senses.

Basic Principles

- **Design to control air movement** - Limiting & controlling air movement is critical because sound needs air movement to travel. The airtightness of a building or facade can achieve control of air movement and consequentially sound movement through the following:

    - **Separation** – being far enough away from the source of the sound so that its energy wave passing through the air is depleted.

    - **Barrier** – forming barriers to the air movement created by the sound energy.

    - **Absorption** – absorption of the air movement caused by the sound energy.

Design occurs at various scales of operation and at different stages in the design process, and the designer can choose the most efficient tactic to control air movement, and consequentially noise depending on the stage in the design process. The scale of operation a designer is working at can range from the following:

> **Masterplanning / Large site planning** – focused on the layout and position of buildings on a site and the urban-form created by the positioning of the building blocks. At its extreme this can include the design and planning of entire new settlements.
>
> **Internal space planning / room layout** – Once the external perimeter of the building and its position within a site has been decided, or if the building is an existing building that is being adapted, then the designers focus is primarily on the location of spaces within the building envelope, and room layout.
>
> **Building components & acoustic treatments** – Once the layout of the buildings on site, urban form, and internal space planning has been decided on, the final scale of operation for the designer is detail design. The design of the details of the building is primarily focused on the components (e.g., windows) materials and where they join together, which could include acoustic treatments, or when choosing the composite of materials that will form the enclosure of the building.

At all scales of operation designers can create designs that can control sound energy.

The design solution for noise control will also depend on whether the development is a refurbishment, an infill development, a new precinct, or a new settlement. Understanding context, noise level, and where the noise is emanating from allows the designer to choose an initial appropriate tactic for the scale of operation and the specific stage in the design process (i.e., strategic/master planning/site layout, building design, or detailed design). Depending on the context and the stage of design, this could be creating a perimeter *street-wall* as a barrier to the noise with a quieter courtyard to the rear, and designing the internal space-planning to respond to the noise sources (see Figure 94 and Figure 103). Alternatively, separation from the noise source may be an option. As the design progresses to a more detailed stage to look at the interior space planning of the building, rooms can be oriented so that noise-sensitive uses/spaces are further away from the noise source (see Figure 100 to Figure 103).

Finding a solution to noise from a busy road may require a different design approach from a solution attempting to reduce aircraft noise, because of the composition of the noise and the direction it is emanating from, and what is appropriate in the Settlement and Habitation context. A suggested hierarchy of design response is:

		Potential costs	Potential development type
1	Separation	Cheaper	New build
2	Master-planning/Urban Form/Building Layout on Site		
3	Internal Space Planning/Room Layout		
4	Acoustic Treatments		
5	Acoustic Devices that also provide ventilation	Expensive	Existing

4.1 THE DESIGN PROCESS

Considering noise from the outset

The creation of a design usually changes through iteration and progresses to an optimal solution for a given set of constraints. The opportunities & constraints will include the site and broader context and the client's requirements. The need to design for the specific characteristics of a context is not new to Architecture, and it is even noted in De Architectura by Vitruvius in book VI, Chapter 1, Paragraph 1, written about 27BC. Architects and designers must carefully analyse the opportunities & constraints of a site in conjunction with the needs of their client(s). Some opportunities & constraints may have already been identified in planning-controls and can be a starting point for the architect, or designer. Planning control maps like those discussed in the previous section *1.8 Acoustic Mapping,* are an excellent way to highlight noise-prone areas. However, an architect or designer should use their common sense and design for noise if it is evident that noise could be a problem. Noise is often overlooked in site analyses and failure to produce a thorough site-analysis can lead to additional costs and delays for a client later in the design process.

Several systems seek to break down the design process into stages, which are mainly produced by professional bodies including the following;

> The Royal Institute of British Architects Plan-of-Work has eight stages,

The American Institute of Architects has seven stages,

The New Zealand Institute of Architects also has seven stages,

The Australian Institute of Architects has six stages, and several project management systems have five stages.

Whichever system you are using it is essential to attend to any noise issues in the early stages of design because it may cost your client more in the later stages. In the early stages of design there is more opportunity to design-out some forms of noise through the positioning of buildings on a site, and through the internal layout of rooms and it is prudent to address this before seeking planning approval. If a designer leaves solving noise problems to the later stages of design, they may have to rely more heavily on expensive components and sound absorption as solutions. When beginning the design process for a noise affected site, it may be better to consider that there is a combination of solutions to mitigate noise, rather than a specific solution for a type of context, or a variety of noise.

4.2 THE SPECTRUM OF TACTICS

One size does not fit all

Rather than try to prescribe one tactic or solution, it is more useful to think of a 'spectrum of tactics' to solve noise issues.

Figure 16 illustrates the spectrum of tactics for various noise sources. The noise-solution tactic deployed will depend on the context, the underlying economics of land value in a settlement, and the type of development (e.g., refurbishment, in-fill development, large new development precinct).

The settlement and habitation transect may help in categorising the context of a project and aid understanding of the tactic to use.

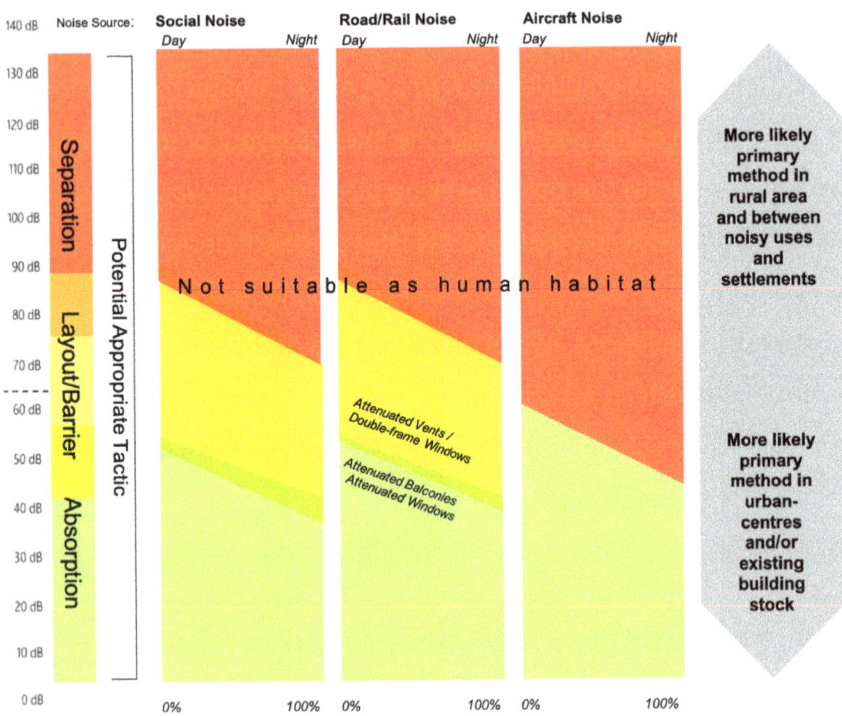

Figure 16 – A rough guide to façade decibel level and possible solutions; by J.T. Cooke with assistance from Brian Cox; July 2018 © J.T. Cooke. *Note: at lower decibel levels and higher frequencies, a combination of absorption and layout changes to create a barrier may be more effective in reducing noise levels. As decibel levels increase, separation in combination with layout changes is the most effective solution, until decibel levels reach levels where separation is required, and noise sensitive uses should not be proposed. Disclaimer: This is a rough guide and you should seek the guidance of a qualified acoustician for your project, The publisher and the author make no representations or warranties with respect to the accuracy or completeness of the content of this work and specifically disclaim all warranties of fitness for a particular purpose.*

There are many solutions to noise, but most are combinations of three tactics:

Separation - merely moving away from the noise because noise levels reduce with distance from the source. Less noise-sensitive uses could be built in the separation distance, but then it is a barrier tactic.

Making a barrier to the noise - this could be a separate barrier structure, designing the site layout so that it forms a barrier, and designing the space-planning and layout of the building to create a barrier.

Absorption & diffusion – this is primarily the use of materials, building elements and devices to mitigate noise.

As discussed previously, some jurisdictions provide sound level heat-maps at a strategic planning level. The strategic planning sound level heat-maps are very useful to alert designers and planners to the presence of a particularly noisy acoustic context. The identification of a noisy acoustic context can help prompt developers and the design team to appoint a qualified acoustician at the beginning of the design process, and prevent abortive design work. The acoustician can measure noise levels and provide more detailed heat-maps to make the invisible acoustic context visible, and model the effectiveness of different designs. At the beginning of the design process, designers should measure the facade decibel level in the context of their proposal to get a feel for the levels they will be working with. The *Rough Guide; Noise Level and Spectrum of Tactics* in Figure 16 could be of help to choose the most likely potential primary tactic. However, the advice of a qualified acoustician must be sought. The development's position in the settlement and habitation transect will also be a consideration in selecting the primary tactic. Separation from the noise source will be more likely in a new-build, rural context because the land values may make separation cheaper than a more complex layout, or absorption component response, which can be costlier. Generally, where decibel levels are low and sound wavelength short, then absorptive materials can be used to reduce the noise to acceptable levels, and designers should use common sense in assessing the response to context.

4.3 MATERIALS

Considering noise when specifying materials

Architects and designers don't always consider the impact their choice of material has on the acoustic context. The choice of materials and finishes in and on a building is critical in creating an excellent acoustic context and good human habitat. Many of you will have experienced being in a restaurant or bar where the lack of soft furnishings has made the din of noise and reverberation unbearable. Architects and designers sometimes consider the impact the choice of material has on noise in internal spaces; however, architects and designers seldom consider the effect the choice of material has on external spaces. Understandably, robust weatherproof materials are usually specified for external spaces, and one unfortunate consequence can be reflected noise. Several factors influence the choice of materials, including; the context, the environment & weather, if the location is internal or external, the level of robustness and permanence required, the aesthetics required, and the cost. Equally important is how the material choice will improve or exacerbate noise, but often robustness, aesthetics, and price, take primacy over acoustic considerations.

As discussed in *Chapter 1 The Phenomena of Sound* the pressure waves we perceive as sound need a medium, or matter to transmit its energy. In the case of airborne sound, the medium, or matter is air. In the case of other material matter, the sound energy's behaviour depends on the specific acoustic qualities of the material. Materials are just another form of matter, and when a sound wave is transmitted through the air and meets another form of matter, such as a concrete wall, the behaviour of the sound wave energy will change. The energy is not lost, and depending on the properties of the material, some of it could be reflected, or transformed into another form of energy, such as heat. Materials can reflect, absorbs and transmits sound, and the specific properties of the material matter determine how the sound energy will behave. A distinction should be made between the inherent acoustic properties of a material and additional properties of modified materials.

In addition to the inherent acoustic qualities of a material, modifying the shape and form of the material can improve its acoustic performance further. One example is timber, and wood varies from denser hardwood to more porous softwood. Wood can be an excellent material for musical instruments because of its resonant acoustic qualities, but it isn't particularly good at absorbing sound. Denser hardwood is less absorbent than softwood, and holes can be drilled in the timber panel to improve its absorptive qualities. The inherent acoustic qualities of common building materials are listed below:

Material	Primary Acoustic Strength	Primary Acoustic Weakness	Material Density (High to Low)	Fibres & Molecules ability to move and transform sound energy to heat
Steel	Reflecting sound	Transmits sound from impact	High	No movement
Concrete	Reflecting sound	Transmits sound from impact	High	No movement
Brick (wall)	Reflecting sound	Transmits sound from impact	High	No Movement
Concrete Block Wall	Reflecting sound	Transmits sound from impact	High	No movement
Glass	Reflecting sound	Transmits sound from impact	High	No movement
Mineral Fibre	Absorption	Does not reflect sound. A great thickness may be required for long wavelengths	Low	High movement
Fabric	Absorption	Does not reflect sound. Great thickness is required for long wavelengths	Low	High movement

FIGURE 17 – RECYCLED GLASS SOUND ABSORBING PANEL, FROM QUIETSTONE; ACOUSTIC PANEL WALL SYSTEM, 2018 RETRIEVED FROM WWW.QUIETSTONE.CO.UK; MARCH 2018. © QUIETSTONE

The material can change the sound wave energy, and the requirements of the acoustic context should, in part, determine the choice of material. The choice of material can change the sound wave energy primarily by absorption, reflection, and transmission of the energy. Dispersion, refraction, diffraction, and combinations of all phenomena also alter the sound energy. If the design objective is to reflect sound waves, denser materials such as concrete perform better than less dense materials such as mineral wool. However, if one is seeking to absorb sound wave energy and convert it to heat, then mineral wool is far better than concrete. Some materials are designed to have robustness and sound absorption qualities, materials such as Quietstone use glass fibres to absorb sound, see Figure 17.

FIGURE 18 – COLOURFUL FABRIC ABSORBER, REPRINTED FROM ARTNOVION AZORES RANGE, 2018, RETRIEVED FROM WWW.ARTNOVION.COM; SEPTEMBER 2018. © ARTNOVION

The size of the obstruction compared to the size of the sound wave will also determine how the wave will behave. We discussed and saw how waves behave when diffracted around obstacles in *Chapter 1 The Phenomena of Sound*. Although it is not precisely the same, it may be useful to think again about waves on the sea. The seawall reflects waves smaller than the barrier, but when a wave is larger than the barrier, it will burst over the seawall. Keeping with the sea-side analogy, if a wave meets sand or pebbles, some of its energy is absorbed in moving the sand and stones. The wave energy is converted to kinetic energy moving the pebbles, similarly with fibre absorbers; the sound energy is converted to kinetic energy and heat by the sound-waves movement of the fibres. Fabric absorbers are generally for indoor use and can be aesthetically pleasing and striking designs like one illustrated in Figure 18. The thickness of the material will also affect the behaviour of sound energy. For example, a very thick concrete floor will perform well at reflecting airborne sound. However, a concrete floor may still transmit sound energy through the concrete. If you have been in a multi-storey building, you may have experienced hearing the sound of people walking on the floor above you. You probably couldn't hear the people talking, because the concrete floor reflected the airborne sound, but transmitted the sound energy caused by the impact of footsteps on the concrete floor. In this type of scenario, the design of the floor requires a composite of materials chosen to isolate impact transmission and reflect airborne sound energy.

4.3.1 COMPOSITES OF MATERIALS

A composite of materials includes several materials chosen for their acoustic properties. In the example of a concrete floor in a multi-storey building, the inclusion of carpet, or other soft absorptive material would cushion the impact of footsteps and inhibit transmission of the sound energy. Composites of materials are conventional in building, and specifying layers of different material for internal walls to benefit from their combined acoustic performance is standard.

Similarly, floors and roofs are usually a composite of materials chosen for their combined acoustic and aesthetic qualities, and durability. By using a composite of materials, it is possible to combine the sound reflective attributes of a concrete wall with the sound absorption qualities of a fabric and mineral wool acoustic panel.

4.3.2 THE FORM & SHAPE OF MATERIALS

The form and shape of material will also affect its acoustic performance. A continuous brick wall performs well in reflecting sound energy; however, a non-continuous brick wall will let sound through the gaps, see Figure 85 as an example. The size of the wall relative to the wavelength of the sound will also determine its acoustic performance. To reflect long wavelengths a very high, often impracticably high wall may be necessary. Where two sound reflectors are parallel, such as two walls, the space between the walls will be subject to reflected noise and reverberation. Reflected sound and reverberation is typical in dense urban cores, where the high buildings made of concrete, brick and steel reflect sound and reverberate it around the street. The building facade shape and material can affect the quality of the acoustic context of the street, and the soundscape of the area, but designers seldom consider this.

Concert Halls have precise forms & shapes, and their interior shape is honed for optimal acoustic performance. A concave shape focuses and gathers sound energy. Ancient Greek amphitheatres were built of stone, a material that is excellent at reflecting sound. The concave form enclosed and reflected the sound of the orator speaking, and the softness of the audience provided acoustic absorption. The skill and knowledge the ancient architects and designers possessed are incredible, and it is a shame that contemporary designers appear to have lost much of this ancient wisdom. If you have ever stood on the stage of the Ancient Greek amphitheatre at Epidaurus, you may have experienced how the speaker's voice is perfectly intelligible to the audience. Readers can inspect the context of the Amphitheatre at the following coordinates:

37°35'46.5"N 23°04'45.1"E

and the following links:

openstreetmap.org

4.3.3 THE POSITION RELATIVE TO THE NOISE SOURCE

The material's performance is affected by its position relative to the noise source and the receiver (the person). Generally, a noise barrier such as a brick wall will reflect more of the sound energy if it is closer to the noise source, and the receiver will experience less noise if they are on the other side of the wall. Look at Figure 19, if the car making noise is further away from the wall, it will be a less effective barrier to the noise, and if the receivers (people) are closer to the wall, as the couple is, they won't hear the noise as they are in the shadow of the barrier. If you hold your hand close to your mouth and speak, less sound escapes your grasp than if you keep your hand away from your mouth and make the same noise. We will discuss noise barriers later in the book. Similarly, the position of an absorptive material relative to the noise source, and the receiver will affect the acoustic performance.

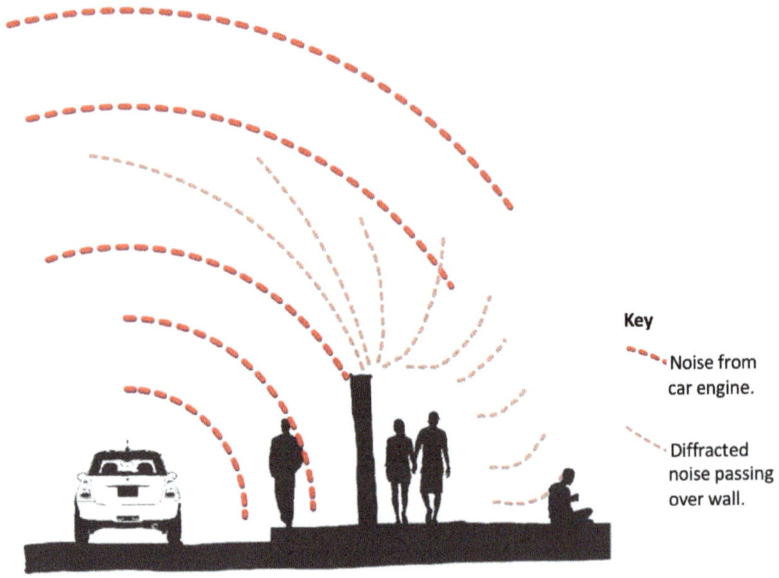

FIGURE 19 - POSITION OF THE RECEIVER RELATIVE TO THE NOISE SOURCE; BY J.T. COOKE; JULY 2018 © J.T.COOKE. *NOTE: THE NOISE SOURCE IS THE CAR ENGINE AND THE PERSON ON THE FOOTPATH FEELS THE FULL FORCE OF THE SOUND WAVE. THE COUPLE WALKING BEHIND THE WALL ARE IN THE SHADOW OF THE WALL AND BECAUSE THEY ARE SO CLOSE TO THE WALL THE SOUND WAVES DO NOT REACH THEM. THE PERSON SITTING ON THE GROUND IS AFFECTED BY SOUND WAVES THAT HAVE BEEN DIFFRACTED (BENT) AROUND THE WALL. CAREFUL DESIGN OF THE TOP OF A BARRIER CAN REDUCE THE AMOUNT OF DIFFRACTED SOUND.*

4.3.4 SURFACE

A rough surface like an irregular stone wall could both reflect and scatter sound waves. Forming timber into an irregular surface like that shown in Figure 27 will scatter and diffuse sound energy. A rougher, or irregular surface on a wall or building facade could also help to reflect and scatter sound wave energy. Shapes that reflect diffuse sound in conjunction with materials that absorb the sound can increase the amount of sound energy absorbed, and diffusers are discussed further in the next section. Sound absorbing and diffusing surfaces &

materials can be both aesthetically pleasing and functional (see Figure 18, Figure 20, Figure 23, and Figure 27 to Figure 28). Still, softer fibrous absorptive materials will not be suitable for areas exposed to the elements and will need protection with another material. Robust stone-like porous materials made from recycled glass are available and can be useful in areas exposed to the elements (see Figure 17).

FIGURE 20 - COLOURFUL FABRIC ABSORBER, FROM ARTNOVION AZORES RANGE, BY ARTNOVION 2018, RETRIEVED FROM WWW.ARTNOVION.COM; SEPTEMBER 2018, © ARTNOVION

As mentioned previously, combining materials for their acoustic qualities to create new materials or composite materials can significantly increase the sound attenuation efficacy. When creating composites, changes to the shape and composition of the surface of a material can also improve acoustic performance. I will discuss surfaces further in the section on reflection and diffusion.

CHAPTER 5

ABSORPTION DIFFRACTION REFLECTION & DIFFUSION

Tactics & Solutions Primarily for the Settlement-Centre, & Existing Building Stock

FIGURE 21 - THE SETTLEMENT CENTRE, AND CENTRE-EDGE; BY J.T. COOKE; NOV 2019 © J.T.COOKE.
NOTE: THE CENTRE OF SETTLEMENTS IS TYPICALLY THE OLDEST PART AND CONSISTS PRIMARILY OF EXISTING BUILDING STOCK, WHICH MAY HAVE HERITAGE SIGNIFICANCE. THE TYPES OF SOLUTION IN CHAPTER 5 ARE PARTICULARLY USEFUL IN THIS TYPE OF CONTEXT. THE CENTRE-EDGE OFTEN HAS RING-ROADS OR OLDER ARTERIAL ROUTES, AND TRAFFIC-RELATED NOISE ISSUES MAY HAVE INCREASED AS THE SETTLEMENT GREW, AND TRAFFIC VOLUMES INCREASED. HOUSING, SCHOOLS AND OTHER SENSITIVE USES LOCATED ON ARTERIAL ROUTES THAT WERE ONCE QUIET ROAD WITH LESS TRAFFIC CAN BE MORE AFFECTED BY NOISE BECAUSE OF INCREASED TRAFFIC VOLUMES.

Generally, the urban-centre of a settlement consists of the older building stock and is denser than areas in the suburbs and the periphery of a settlement. For these reasons, it is more likely that there will be insufficient space to use separation alone to control noise. Working on existing and older building stock means that refurbishment is more likely and that significant changes to the space-planning of the building are less likely. Replacement of windows with double-frame windows, or acoustically attenuated fresh air ventilators, or other component solutions are more likely tactics when improving existing building stock, because of the limited opportunity to reconfigure space planning or change the building's layout on site. In dense historical cores, social noise, and nuisance noise is more likely to be the cause of sleep disturbance. As illustrated in section 3, in many instances' railway lines penetrate deep into the urban core of cities and often end in a central terminus station, or stations, and create a corridor of noise-compromised buildings along the railway route, see Figure 11. Developing in urban cores along these noise corridors can be challenging and it may be that re-configuration of a building's layout, or forming a barrier is the required to create good human habitat.

5.1 ABSORPTION

Absorbing and converting sound energy

An absorption solution to mitigate airborne noise is more likely to be specified at the detailed design stage compared to layout/barrier and separation, and is easier to incorporate at later design stages. Absorption in combination with intelligent design to reflect, diffuse and diffract most of the noise can provide an effective solution. Depending on the source of the noise and its path of entry into the internal space, absorption materials and resonators can reduce airborne noise before it enters the internal space.

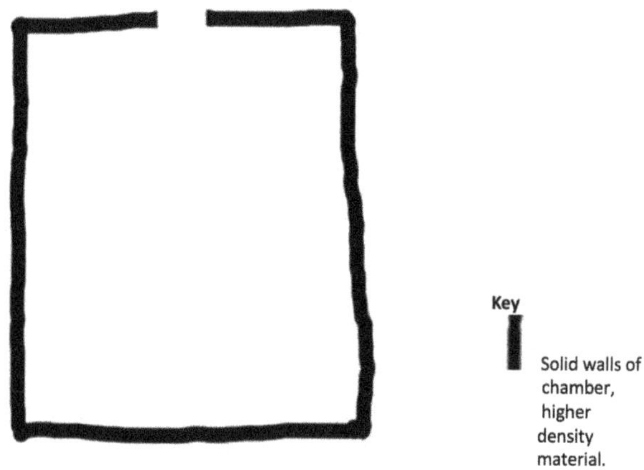

Figure 22 – A section through a Helmholtz type resonator, by J. T. Cooke; June 2018 © J.T.Cooke. *Note: A Helmholtz type resonator can be any enclosed volume of air with a narrow opening, a simple glass bottle can be used as a resonator. Resonators are particularly useful where space is limited.*

Absorbers convert sound energy into heat through mechanical work. Sound absorbers are materials and devices that convert sound energy into heat. Softer materials that allow movement such as mineral wool is better at absorbing sound wave energy than dense materials. Devices, and the particular shape of the material can help absorb sound energy depending on the frequency of the sound energy. One type of absorber illustrated in Figure 22 is a device called a Helmholtz resonator, which traps a piston of air at its narrow opening and vibrates it back and forth, converting the sound energy to heat, see Figure 26. If you have ever noticed the little holes in acoustic timber panels like those shown in Figure 23 these behave in the same way, albeit usually with soft material behind them. The soft material behind the timber panel usually absorbs some higher frequency sound energy. The pistons of air in the holes in the panels that act as resonators can be spaced and sized (tuned) to absorb specific frequencies.

The predominant weather characteristics and climate zone will affect how sound behaves and, consequently, the effectiveness of the material. Generally, in hot and humid weather, absorption can be more effective because the speed of sound increases in warm, moist air and absorptive materials perform better as the speed of sound increases. However, designers must take care to prevent some fibrous materials from becoming damp.

All absorbers convert sound energy into heat through mechanical work and movement. The conversion depletes the sound energy transforming some of it to heat. There are broadly three main types of absorber:

> **Fibrous absorbers –** These are compressible fibrous materials with air spaces between the fibres. The sound energy causes the fibres to vibrate and transforms sound energy into heat, depleting the sound energy. See Figure 24
>
> **Mechanical absorbers –** These are something that takes sound energy and converts it to movement (kinetic energy) and heat

(thermal energy). A panel mounted on springs could be a mechanical absorber. See Figure 25

Resonance absorbers – A glass bottle is the simplest resonator. The bottle shape combined with a specific frequency of sound traps a piston of air in the bottle opening. The vibration of the piston of air molecules transforms the sound energy to heat and movement depleting the sound energy. See Figure 26

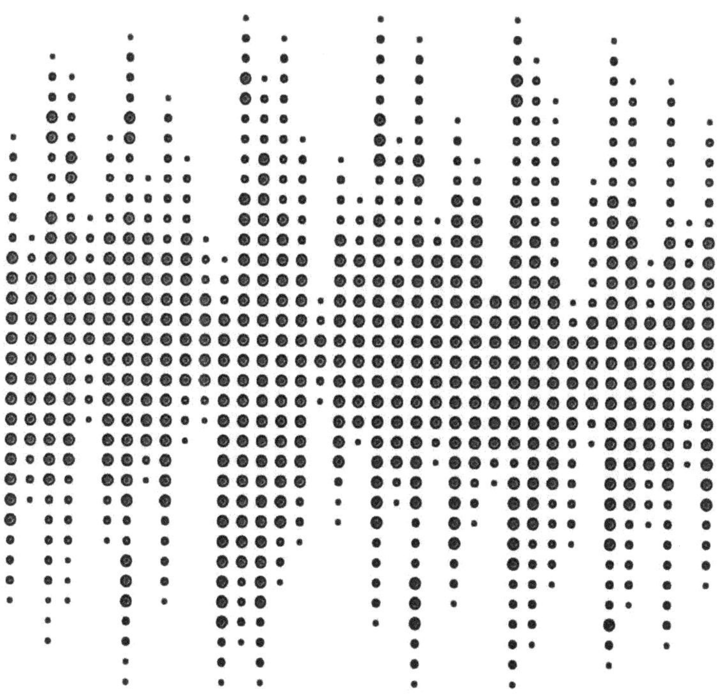

FIGURE 23 – PERFORATED TILE ABSORBER PANEL, FROM ARTNOVION ATHOS W RANGE; BY ARTNOVION, 2018, RETRIEVED FROM WWW.ARTNOVION.COM SEPTEMBER 2018. © ARTNOVION

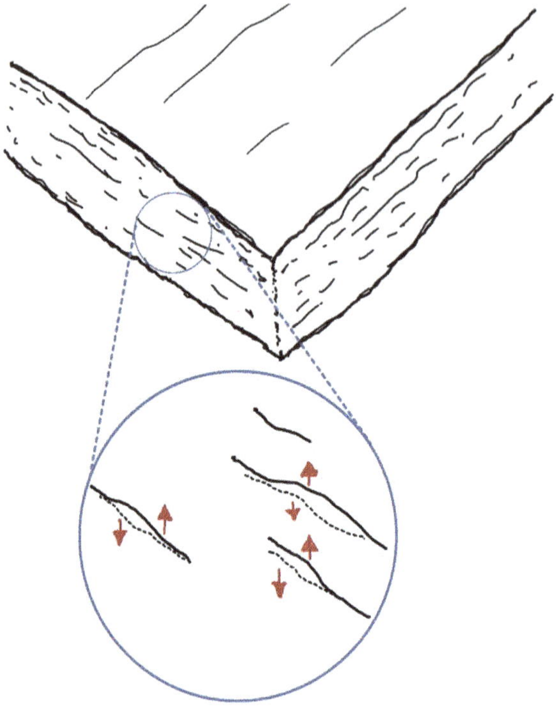

Figure 24 – A Mineral wool absorber; by J.T. Cooke; Nov 2019 © J.T.Cooke. *Note: The sound energy is converted to kinetic energy through the movement of the fibres. The movement of the fibres converts some of the kinetic energy to heat, which is dissipated in the surroundings. This process lessens the sound energy.*

Architects and designers will probably be familiar with perforated and fibrous absorber panels, like those illustrated in Figure 23 and Figure 24. However, they are likely less familiar with resonators. Architects & designers will be familiar with using panel absorbers for internal space. Still, it may be possible to use some absorption materials and devices like resonators externally where the weather allows. Resonators may be a more robust option where the climate precludes the use of some fibrous absorbers.

The angle of the source of the noise to the absorptive material also affects the effectiveness of the absorber. Thinking again of energy

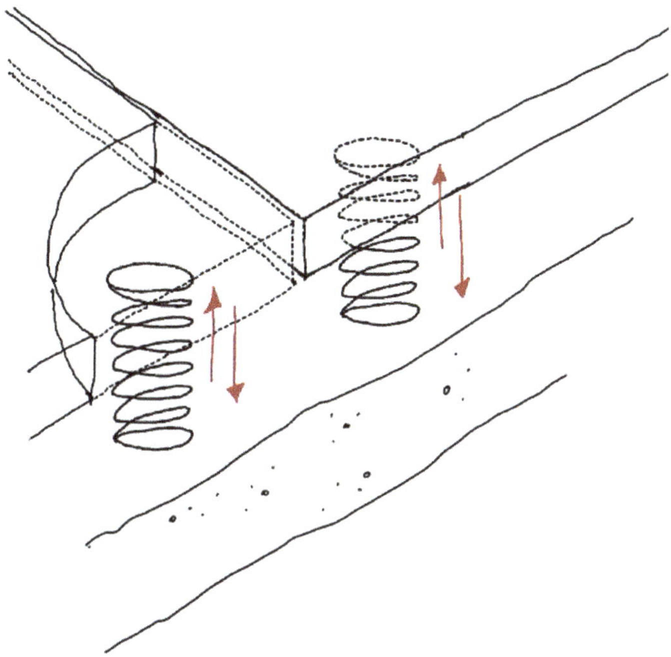

Figure 25 - Mechanical panel absorber; by J.T. Cooke; Nov 2019 © J.T.Cooke. *Note: The sound energy is converted to kinetic energy through the movement of the springs. The movement of the springs converts some of the kinetic energy to heat, which is dissipated in the surroundings.*

instead of sound, if the energy hits the absorbent material at 90 degrees to the absorptive panel the energy is more likely to penetrate the fibres of the material and cause them to move and so convert the sound energy into heat. Sound energy hitting an absorptive panel at an oblique angle is more likely to have some of that energy reflected into the environment. If you can imagine a snooker table, and if you hit the white ball (cue-ball) very hard against a red ball (or any ball), the cue ball will stop. The ball hit will move, and this is because the cue ball has transferred all of its energy to the other ball, however, if the cue ball hits the red ball at an obtuse angle (not direct on) then both balls move. Both balls move because the angle of incidence didn't allow for the cue ball to transfer all its energy to the red ball, and this is similar to sound energy hitting an absorption panel.

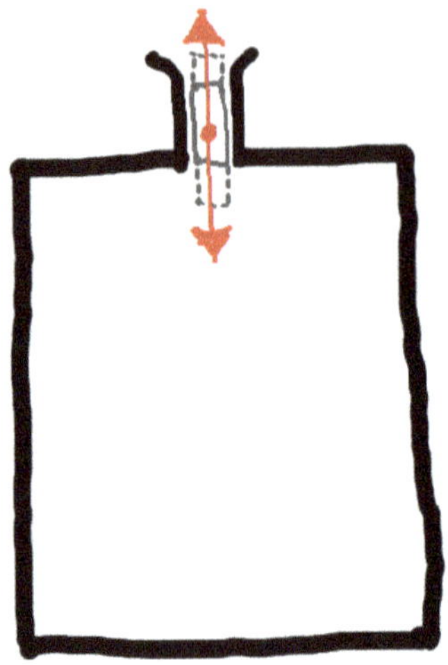

Figure 26 - Resonance absorber or Helmholtz absorber; by J.T. Cooke; Nov 2019 © J.T.Cooke. *Note: The sound energy is converted to kinetic energy through the movement of a section of air in the neck of the resonator. The movement of the air converts some of the kinetic energy to heat, which is dissipated in the surroundings.*

A rating between 0 and 1, called the absorption coefficient classifies absorptive materials. The material is classed 1 if all energy is absorbed (the cue ball stops), and 0 if nothing is absorbed (you missed the red ball).

In addition to the angle of the absorber, the position of the absorber relative to the noise source is important in optimising the materials sound absorption. The optimal location of the absorber is related to the individual wavelengths of the sound one is trying to absorb. If targeting the absorption of several wavelengths, the optimal position for one wavelength may be in a different location than for another wavelength in the sound you are trying to absorb. Designers can choose to have a thickness of absorber that occupies the optimal position for several wavelengths or have separate absorbers in the different optimal locations. Designers can use the Quarter Wavelength Rule to calculate the optimal position for an absorber from a noise source.

5.2 THE QUARTER-WAVELENGTH RULE

A rule-of-thumb for positioning absorbent material

A useful rule to calculate the optimal position of an absorbent material such as mineral-wool is the Quarter Wavelength Rule. The position is optimal for a specific wavelength and if you are trying to absorb a broad spectrum of sound (many wavelengths) the Quarter Wavelength Rule will mean there will be a different optimal position for the different wavelengths, or you can have a sufficiently thick section of absorbent material so that it is able to occupy the optimal position for the range of wavelengths being absorbed. For example:

$$Optimal\,position\,of\,absorber = \frac{wavelength}{4}$$

Remembering the formula from Chapter 1, with a wavelength of 30Hz, and an assumed dry air temperature of 20° and standard atmospheric pressure, then the wavelength in metres is as follows:

$$11.43\ \text{m} = \frac{343\ m/s}{30Hz}$$

Now that you know the wavelength in metres you can simply divide by four as follows:

$$2.86m\ is\ the\ optimal\ position\ of\ the\ absorber = \frac{11.43}{4}$$

As discussed in Chapter 1, the atmospheric pressure, the temperature, and level of humidity can all affect the resultant wavelength and therefore optimal position of an absorber, and it is prudent to cover a range of positions. This calculation is for just one wavelength of sound and you may be targeting a broad spectrum of sound waves for absorption, and therefore a greater thickness of material to occupy all the Quarter Wavelength positions may be prudent. The method adopted of having individual absorbers at the various Quarter Wavelength positions, or having a continuous greater thickness of material will be determined by the specific wavelengths being absorbed and the cost of material. Taking the worked example, rather than having 2.86m of absorbent material it would be more efficient to mount the absorbent material 2.86m from the wall, or ceiling, facing the sound waves to be absorbed.

In the section on materials, we discussed the positioning of materials and noted how composites of materials were functional when trying to mitigate different types of sound transfer. Similarly, we discussed absorbing longer wavelength sound waves; the absorbent material can be placed at approximately one-quarter of the wavelength of the sound wave, and it will have a similar level of performance than if it were filled with absorbent material for the total distance. It may be prudent to coordinate the required absorptive material positions with your material composite build-ups.

5.3 DIFFRACTION

Bending sound waves

It can be challenging to design out diffracted sound, and it can pass over the top of barriers or around them. Diffracted sound tends to be low frequency (longer wavelength) and can be difficult to reduce by absorption (U.S Department of Transportation; Federal Highway Administration, 2017). Figure 19 illustrates how diffracted sound waves can pass over a barrier. The person sitting behind the wall is receiving car engine noise diffracted over the barrier wall. The design and material choice of the top & edge of a wall can help mitigate diffracted sound. Careful detailing to create barrier tops in a particular shape made of absorbent material can reduce diffracted sound waves, see Figure 35-D, and section *6.1 Noise Barrier Structures* for more on designing barriers. It may be helpful to watch the video at the QR code link for further explanation.

YouTube Channel

5.4 REFLECTION & DIFFUSION

Scattering sound waves

Sound waves can reflect off a barrier. When a sound wave hits an obstacle, if the obstruction is large enough relative to the size of the sound wavelength, it will become a barrier to the sound energy and reflect it. The material composition of the obstruction is also important, and denser materials like concrete are better at reflecting sound energy than absorbent materials. Again, it may be helpful to think of waves in the sea and a sea wall. If the sea wall is large enough relative to the size of the sea wave, the wave will reflect off the sea wall. If the sea wave is higher than the sea wall, it will crash over it. Similarly, soundwaves with a long wavelength may pass over a barrier if it is too small compared to the size of the wavelength of the soundwave. It may be helpful to watch the video at the QR code link at the end of this section for further explanation.

Diffusers do not remove sound energy but spread it more evenly, and can be used in conjunction with, or as an alternative to absorbers. There are many different types of diffuser, but they all consist of different lengths of material that vary the position of the sound reflective surface, scattering the reflected soundwaves. A diffuser can be used in conjunction with sound absorbent material and form larger & smaller chambers to diffuse and absorb the sound energy.

The Schroeder diffuser is a type of diffuser that has multiple chambers of different depths that can channel, control and change sound waves.

Diffusers can be thought of like floodwater channels; they cannot stop the flow of water, but they can control and change it. The material and design of diffusers can be aesthetically pleasing, and Figure 27 to Figure 28 show some aesthetically pleasing acoustic treatments.

FIGURE 27 - COLOURFUL WOODEN MODULAR HEMISPHERICAL DIFFUSER, REPRINTED FROM ARTNOVION ALPS W RANGE; BY ARTNOVION, 2018, RETRIEVED FROM WWW.ARTNOVION.COM; SEPTEMBER 2018. © ARTNOVION. *NOTE: THE VARYING DEPTH OF CHAMBER CREATED BY THE SQUARE SECTIONS DIFFUSE DIFFERENT WAVELENGTHS OF SOUND ENERGY.*

Diffusers may be a practical option for use outdoors because they can be formed of robust materials including concrete, metal, timber, and plastics. The shape and formation of the chambers in a diffuser are key, and it is possible to form Schroder type diffusers out of concrete or timber. Creating aesthetically pleasing external walls and soffits may be possible outside, where the elements may prohibit the use of some soft, absorbent materials. Designers could form the surface of off-form and pre-cast concrete walls to make a diffuser, and the soffits of roofs and awnings could include timber or metal diffusers. The sound being mitigated will partially determine the shape of the surface of the diffuser or barrier. The cross-section of the chambers in the Schroder diffuser in Figure 28 is the principle shape in Figure 29 at various sizes. Architects and designers working closely with acousticians could deliver some aesthetically pleasing and functional designs.

FIGURE 28 – SCHRODER TYPE DIFFUSER, FROM ARTNOVION ARVON W DIFFUSER FG FAGUS, BY ARTNOVION 2018, RETRIEVED FROM WWW.ARTNOVION.COM; SEPTEMBER 2018, © ARTNOVION.
NOTE: THE VARYING DEPTH OF CHAMBER CREATED BY THE WOODEN SECTIONS DIFFUSE DIFFERENT WAVELENGTHS OF SOUND ENERGY.

For further explanation watch the video at the QR code link below:

YouTube Channel

5.5 SOME PRINCIPLES SHAPES

Shapes to consider when designing for noise reduction

Absorption of sound energy is generally carried out in a chamber, and the absorbent acoustic lining of the chamber performs the function of converting sound energy into heat energy. The chamber captures and contains the noise (sound energy), and the absorbent lining depletes the power from the sound waves converting it to mechanical energy and dissipating it as heat. Make a mental note of the shape of the chamber in Figure 29. Broadly, there are two types of shape used for absorption; the chamber, and the snorkel, which is an elongated version of a chamber (see Figure 30). Readers could equally describe the snorkel as a periscope or dog-leg, but because it is facilitating the movement of air snorkel seems appropriate.

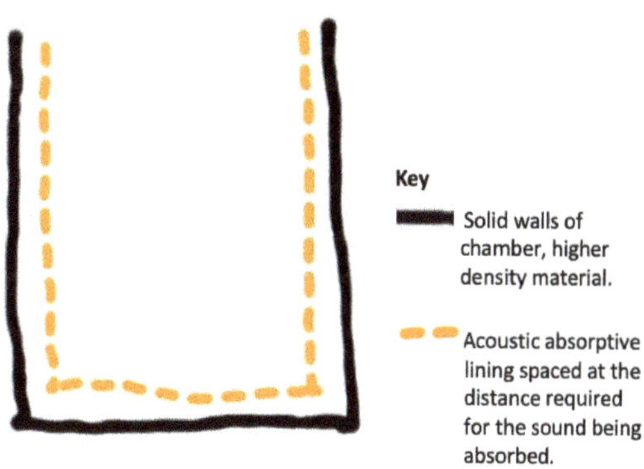

FIGURE 29 – PRINCIPLE SHAPES USED IN ABSORPTION. AN ABSORPTION CHAMBER. BY J.T. COOKE, JUNE 2018 © J.T.COOKE. *NOTE: THE CHAMBER ACTS TO CONTAIN THE SOUND ENERGY AND THE ACOUSTIC ABSORPTIVE MATERIAL LINING CONVERTS THE SOUND ENERGY INTO HEAT, REDUCING THE SOUND ENERGY. WHEN SIZING A CHAMBER, CONSIDER THE WAVELENGTH(S) OF SOUND YOU ARE TRYING TO MITIGATE. THE POSITIONING OF THE ABSORPTIVE MATERIAL SHOULD BE ABOUT A QUARTER TO A HALF THE WAVELENGTH OF THE SOUND BEGIN MITIGATED, SEE THE QUARTER WAVELENGTH RULE. THE DOTTED ORANGE LINE SIGNIFIES AN ABSORPTION MATERIAL.*

In the following chapters we will discuss building components used to reduce noise, and please take note of these primary shapes for absorbing sound energy. Later in the book you will see how the primary shapes in Figure 29, Figure 30, Figure 31 appear in double-frame windows, balconies used as absorption chambers, and sound-absorbing ventilators. In conjunction with some simple techniques in the positioning of ventilation openings, these primary shapes can be the basis of developing some practical solutions to noise, while providing natural ventilation.

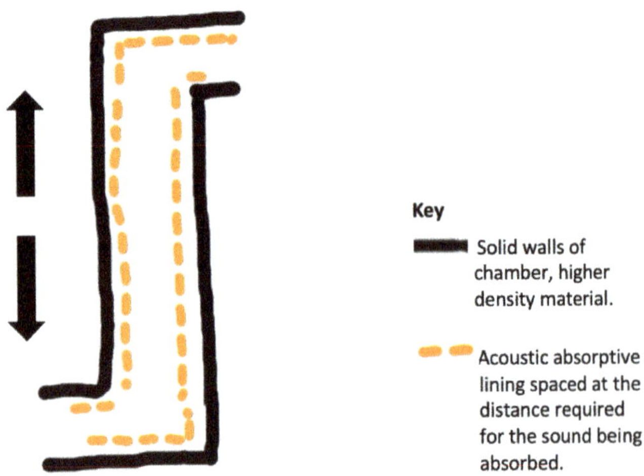

FIGURE 30 – PRINCIPLE SHAPES USED IN ABSORPTION; THE ABSORPTION SNORKEL; BY J.T. COOKE; JUNE 2018 © J.T.COOKE. *NOTE: THIS IS BROADLY ONE TACTIC THE ABSORPTION CHAMBER AT DIFFERENT SCALES, THE ABSORPTION SNORKEL IS JUST AN ELONGATED ABSORPTION CHAMBER. THE LONGER THE SNORKEL OR DEEPER THE CHAMBER, THE MORE RESTRICTED THE AIRFLOW MAYBE AND AN OPTIMUM SHOULD BE FOUND. THE SNORKEL CAN BE ELONGATED TO SUIT THE SOUND WAVES TO BE ABSORBED AS INDICATED BY THE ARROWS. ALSO, CONSIDER THE WAVELENGTH TO BE MITIGATED WHEN SIZING THE CHAMBER. THE DOTTED ORANGE LINE SIGNIFIES AN ABSORPTION MATERIAL.*

Architects and designers can use the principle shapes at different scales and could form a sound absorptive chamber from the articulation in the facade as illustrated in Figure 35 to Figure 37. At the building layout and masterplanning scale, the internal facade of a courtyard form building could be an absorption chamber, see Figure 94. The designer could choose to have absorption chambers of multiple sizes to absorb various wavelengths of sound similar to major (A) and minor (B) absorption chambers illustrated in Figure 35. An open walkway from a noisy area into a courtyard garden could take the form of a maze, as illustrated in Figure 31, and the occupant would experience a change in soundscape when they reach the end of the maze walkway. Working closely with an acoustician from an early stage, designers could coordinate building features such as articulation rebates also to perform a practical acoustic function to reduce noise and integrate the noise solutions into the fabric of the building. Architects could also create different soundscape experiences in different areas in both public and private spaces.

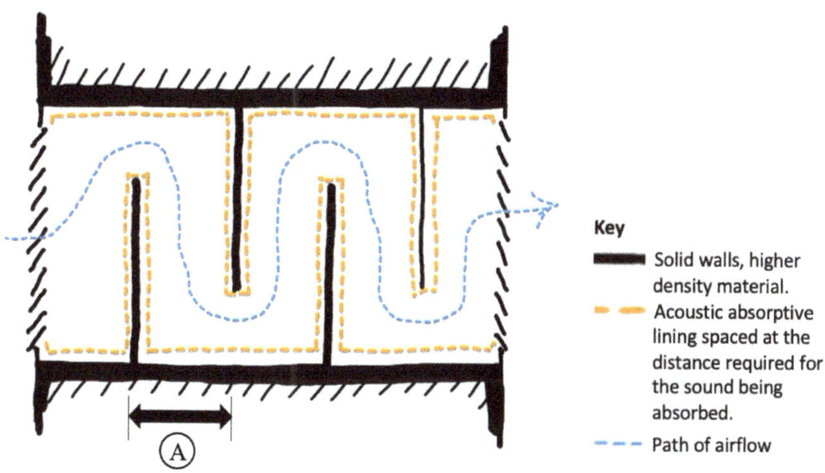

Figure 31 - Principle shapes used in absorption, the absorption sound maze; by J.T. Cooke, June 2018 © J.T.Cooke. *Note: The absorption sound maze can be thought of as a series of snorkels joined together, or a sequence of sound absorption chambers through which the air passes. The greater the number of chambers, the more sound-energy can be absorbed, but the more restricted the airflow and an optimum should be found. Sound mazes are common in ventilation systems. The size of each chamber can be increased or decreased to suit the sound waves being absorbed (see A). The dotted orange line signifies an absorption material.*

5.6 LOCATING VENTILATION OPENINGS & WINDOWS

Understanding how careful location of ventilation openings can reduce noise ingress

Basic Principles

There are some simple things architects and designers can do to mitigate noise, but in the author's experience they do not seem to be widely understood, nor used. The basic principles when locating ventilation openings are the same as when locating a building on a site:

- **Separation** – moving the ventilation opening as far from the source of the noise as possible.
- **Layout/barrier** – making a barrier between the noise and ventilation opening, or rotating the ventilation opening 90 degrees to shield it from the noise source.
- **Absorption** – putting the ventilation opening in a sound-absorptive chamber surrounded by sound-absorptive material
- **Combinations** of all three.

Rotating the ventilation opening at 90 degrees to the noise source is a simple solution to reduce the noise entering the opening, but the author has rarely seen this used for noise control. Waters-Fuller et al. noted in their research, 'Rotation of source incidence away from the normal, within a non-diffuse acoustic environment, is found to consistently improve the resulting open window façade insulation' [3]. (Lurcock, MacKenzie, & Waters-Fuller, 2007, p. 5). The principle here is to use the position of the ventilation opening itself to shield the opening from the noise.

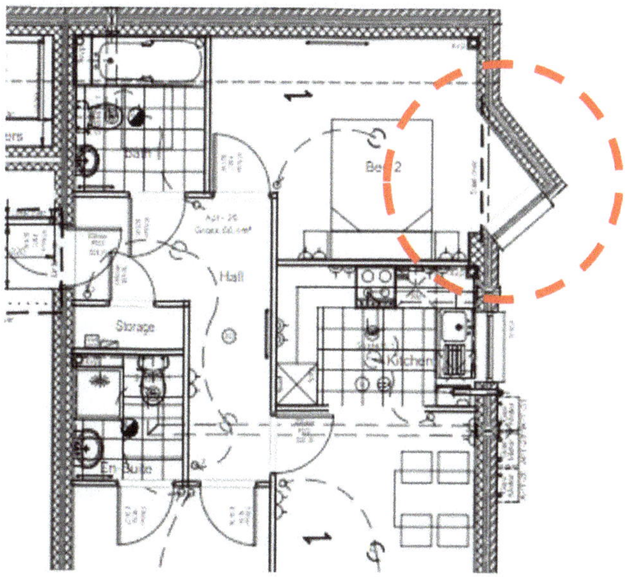

FIGURE 32 – BARRIER BUILDING; BY ECE ARCHITECTURE; IMAGE COURTESY OF ECE ARCHITECTURE; FROM SEVENOAKS COUNCIL WEBSITE; 2015; ELECTRONIC; MAY 2018 © ECE ARCHITECTURE. *NOTE: THE INTERNAL SPACE PLANNING & LAYOUT OF APARTMENTS IN KENT RESPONDS TO THE NOISE SOURCE (A LUMBER YARD).*

It is common practice to use window openings for both natural ventilation and light, and open windows and doors are the primary entry paths for noise. It may help to think of the requirements for human habitat separately: daylight, fresh air, direct sunlight (climate permitting), view and aspect, silence and bustle. It may be that the optimal position for natural ventilation openings is at odds with the best place for daylight, sunlight, or a view.

FIGURE 33 - WINDOWS IN AN APARTMENT BUILDING PROTECTED BY CONCRETE HOODS AND ANGLED AWAY FROM THE NOISE SOURCE; BY J.T. COOKE, MAY 2018 © J.T.COOKE – *NOTE THIS MAY HAVE BEEN DONE PRIMARILY TO PROVIDE VISUAL PRIVACY, BUT COULD ALSO BE USED FOR NOISE REDUCTION.*

Figure 34 - Windows in an apartment building protected by a fabricated hood and angled away from the noise source; by J.T. Cooke, May 2018 © J.T.Cooke – *Note this may have been done primarily to provide visual privacy, but could also be used for noise reduction.*

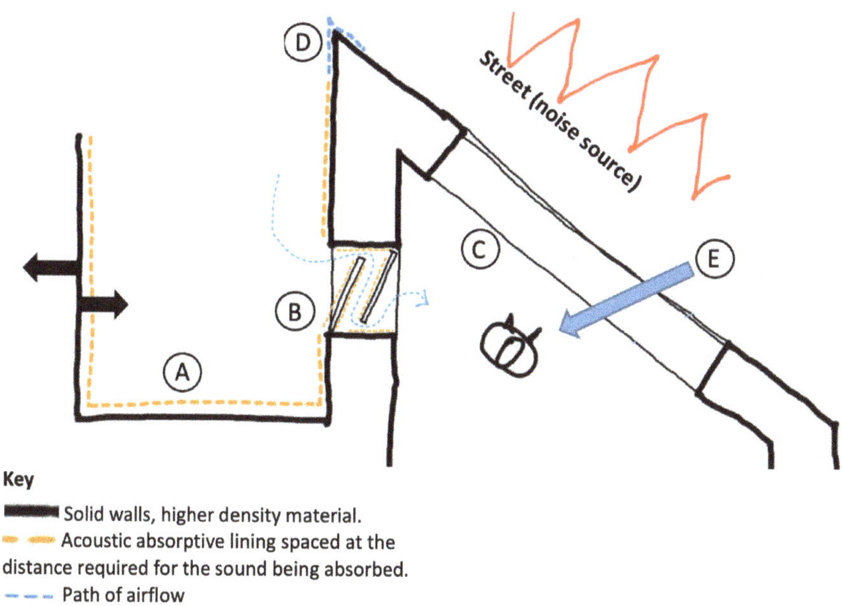

Figure 35 - A typical building situation with windows facing the noise source; By J.T. Cooke; June 2018 © J.T.Cooke. *Note: separating the opening for natural light C from the natural ventilation opening B and placing it at 90° to the noise source in a noise absorption chamber A can reduce noise ingress. The position of the absorption material in the chamber A can be hung to suit the sound wavelength being targeted, as indicated by the arrows. The solid blue line E signifies light, the heavy dotted dark blue line on the corner D signifies a sound absorptive material tip, the absorptive tip can reduce diffracted noise.*

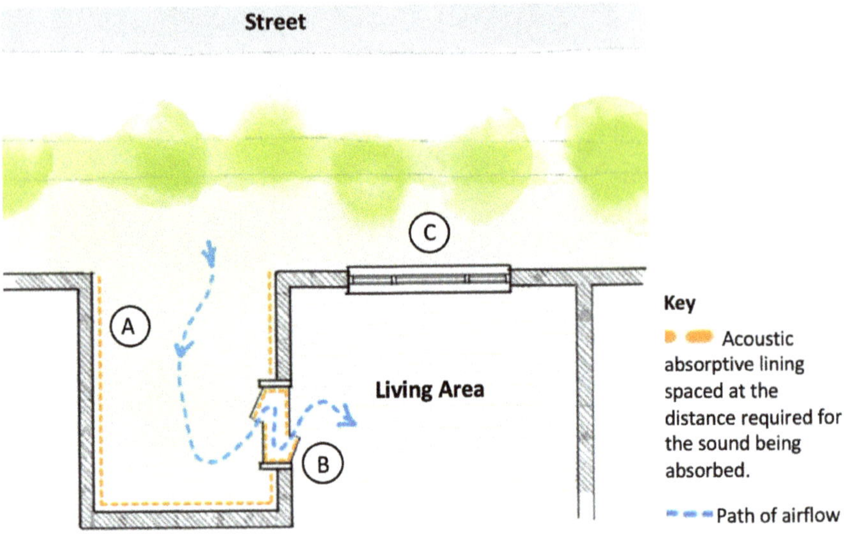

FIGURE 36 - A TYPICAL APARTMENT RELATIONSHIP TO A STREET (NOISE SOURCE) WITH AN INDENTATION IN THE BUILDING LINE; BY J.T. COOKE; JUNE 2018 © J.T. COOKE. *NOTE- A THE BUILDING RECESS CAN BE LINED WITH ABSORPTIVE MATERIAL AND/OR A DIFFUSER TO BECOME AN ABSORPTION CHAMBER. THE POSITION OF THE ABSORPTION MATERIAL CAN BE HUNG TO SUIT THE NOISE WAVELENGTH BEING TARGETED. B THE VENTILATION OPENING IS 90° TO THE NOISE SOURCE AND LINED WITH ABSORPTIVE MATERIAL. C THE WINDOWS FACING THE STREET ARE DOUBLE-GLAZED AND CAN BE SEALED FROM NOISE.*

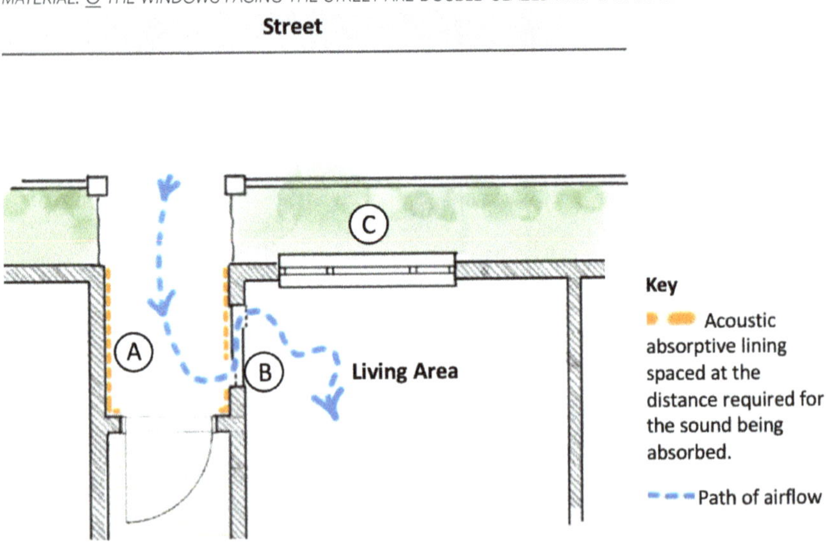

FIGURE 37 - A TYPICAL TRADITIONAL TERRACE HOUSE RELATIONSHIP TO A STREET (NOISE SOURCE) WITH A RECESSED ENTRYWAY; BY J.T. COOKE; JUNE 2018 © J.T. COOKE. *NOTE- A THE BUILDING RECESS CAN BE LINED WITH ABSORPTIVE MATERIAL AND/OR A DIFFUSER TO BECOME AN ABSORPTION CHAMBER. B THE VENTILATION OPENING IS 90° TO THE NOISE SOURCE AND LINED WITH ABSORPTIVE MATERIAL. C THE WINDOWS FACING THE STREET ARE DOUBLE-GLAZED AND CAN BE SEALED FROM NOISE. THIS SOLUTION COULD BE RETROFITTED IN TRADITIONAL TERRACE HOUSING STOCK.*

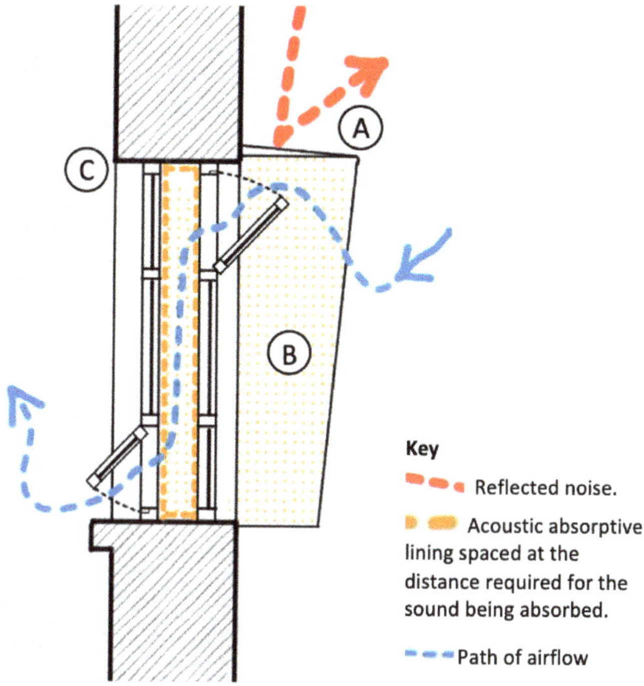

Figure 38 - Absorptive window with noise reflecting & absorbing hood; by J.T. Cooke; June 2018 © J.T. Cooke. *Note- A. Deflective and absorptive hood to reflect noise from above away from the window opening. B absorptive material within hood to absorb noise. C noise absorptive double-frame window.*

Separating openings in the building envelope allows designers to cater specifically for what the aperture is providing to the inside space. If necessary, providing light through a double, or triple-glazed sealed window, and natural ventilation through a separate opening with a sound absorption chamber (see Figure 35 to Figure 38) can maintain acoustic privacy. As previously discussed, in the author's experience, separating openings in the building envelope by function (i.e. light or ventilation), and in combination with orienting the ventilation opening at 90° to the noise source can help shield the aperture from the noise and can provide low-energy, natural ventilation for cooling (see Figure 35 to Figure 38).

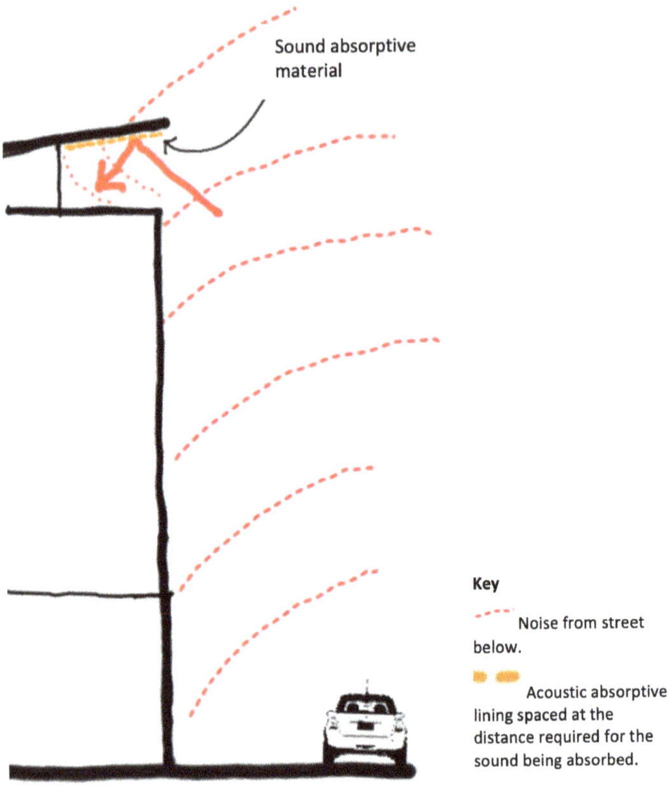

Figure 39 – Section of mono-pitch roof with under soffit acoustic attenuation; by J.T. Cooke; May 2018 © J.T. Cooke. *Note: A common feature of modern apartment buildings is a projecting mono-pitch roof; however, this can exacerbate noise problems and should be carefully designed to include acoustic attenuation. The red arrow signifies reflected sound.*

The acoustic design principle is to reflect or shield the opening from as much noise as possible, leaving the remainder of the noise to be absorbed in a sound absorption chamber. Reduced noise levels are usually possible if the opening for natural ventilation is designed to be at least 90° to the noise source and the building indentation containing the aperture is covered in sound absorptive material (see Figure 35 to Figure 38). The level of noise reduction will need the input of an acoustician and a coordinated acoustic design.

5.7 BUILDING COMPONENTS

Windows and ventilators

Basic Principles

- **Position** - at least 90° to the noise source.
- **Device** - Double-frame window to act as an absorption chamber / snorkel.
- **Separation** - Separation of the opening for light from that of the opening for natural ventilation.

The design decisions at the detailed design stage in a project are usually regarding the specification of building components, such as windows, doors and ventilators. In construction, components are usually items that have been made off-site. The specification of specialist acoustically attenuated components will generally be more expensive than changes to the layout or space-planning. Still, in certain contexts, it may be the only option available to designers. In the previous section, we discussed some principles behind the formation and location of openings and components in a building envelope to help mitigate noise ingress from outside. In this section of the book, we will discuss the design of the components themselves. There have been some incredible advances in acoustically attenuated building components that concurrently provide natural ventilation and some reinvention of historical solutions. Architects, designers and urban planners need to know what solutions are available and in what context they are best used, and they will be discussed and illustrated in the next section.

Figure 40 – An ornate window with a vent below the window in Venice; Photo by J.T. Cooke; May 2007 © J.T. Cooke. *Note: Components such as windows and vents introduced into the building envelope have been made for generations and can be beautiful and functional, as this lovely old example in Venice illustrates.*

FIGURE 41 - TRADITIONAL SHUTTERED WINDOWS IN PORTOFINO, ITALY; PHOTO BY J.T. COOKE; AUG 2016 © J.T. COOKE. *NOTE: THE TRADITIONAL WINDOW AND SHUTTER ARRANGEMENT IN SOUTHERN EUROPE PROVIDES NATURAL VENTILATION WHILE SHADING THE INTERNAL SPACE. THE OCCUPANT CAN TILT THE LOWER LOUVRE PANEL OUTWARD TO INCREASE THE AIR INTAKE WHILE PROVIDING SOLAR SHADING TO REDUCE HEAT GAIN FROM THE SUN. THE LOUVRES WITHIN THE FRAMES CAN ALSO BE TILTED FOR PRIVACY OR TO ALTER LIGHT LEVELS. THE BUILDING OPENINGS IN THESE BEAUTIFUL OLD BUILDINGS IN PORTOFINO ARE ADAPTABLE BY THE OCCUPANTS TO TAILOR THEIR HABITAT TO THE DAILY AND YEARLY CHANGES IN THE WEATHER.*

5.7.1 WINDOWS

Windows are now ubiquitous in buildings, but a summary of the history of windows may help illustrate why we have arrived where we are today regarding window design.

FIGURE 43 - A CONTEMPORARY WINDOW WITH SIDE OPENING VENTILATION SECTION IN PORVOO (BORGÅ) FINLAND; BY J.T. COOKE; JUNE 2013 © J.T. COOKE. *NOTE: THE RESTRICTED OPENING SIZE OF THIS TYPE OF WINDOW CAN CAUSE OVERHEATING IN THE SUMMER MONTHS.*

FIGURE 42 - TRADITIONAL DOUBLE-FRAME WINDOW IN PORVOO (BORGÅ) FINLAND; PHOTO BY J.T. COOKE; JUNE 2013 © J.T. COOKE.

For anyone that has visited an ancient Celtic roundhouse, an ancient Norse longhouse, the need for additional light and ventilation is evident. These old homes were centred around a fire as a source of both heat and light and were smoky. At some stage, a bright spark put a hole in the wall to let in light, ventilation and, unfortunately, noise. The Romans and Egyptians gave us the first glass, and for wetter climates and during winter, the glass windows provided light without letting in the wind and rain. Initially, manufacturers could produce only small panes of glass, and the craftsmen stuck the small pieces of glass together with leading to fill larger window openings. As glass production technology improved, glass panes became larger and cheaper to produce. The window opening continued to be used for both light and ventilation, and the problem of noise entering while windows were open for ventilation continued. In colder climates, windows changed from single panes of glass in a frame to a double-frame window with a single pane of glass in each frame, and later, double-glazing to improve the thermal performance of windows. Double-glazing also enhances the acoustic performance of windows when the window is closed with an air-tight seal; however, once the window is opened for ventilation, the acoustic performance is substantially diminished.

Increasing the layers of glass improves thermal performance, and triple-glazed units are used in very cold climates. The increased layers of glass also improve the acoustic performance while the windows are sealed shut. In Scandinavia, Germany, Russia, Alpine regions, and other areas of Europe that have severe winters, a double-framed window, each frame containing a double-glazed unit, is commonly specified for thermal performance. Figure 44 to Figure 46 illustrates traditional double-frame windows in use in Prague and Finland. The use of double-framed windows has an interesting history, and its use has led to exciting innovations in reducing noise ingress when windows are open for ventilation, which I will discuss in the following section.

The author has read many acoustic reports that assume the sound insulation of an open window as a 10-15 dBA reduction. However, a study conducted at Edinburgh Napier University for the UK Department for Environment, Food and Rural Affairs (DEFRA) by Walters-Fuller et. al. (2007) illustrated that the actual reduction varied widely. The study measured the sound insulation provided by seven different windows with twelve different opening styles. The study found that for a free open area of 0.05 m2 the measured insulation ratings ranged between DW 14 and 20dB, and translated to dBA level differences of:

Road traffic noise 12 – 18 dBA

Railway noise 12 – 18 dBA

Aircraft noise 14 – 19 dBA

Amplified music 15 – 20 dBA

The study found that there was little variation because of the window style, nor the material. The study also investigated closed windows with a 4000mm2 slot and found that the ventilator reduced the performance of the window by 6dB when closed and 11dB when open; therefore, a considerable decrease in performance [4] (Lurcock, MacKenzie, & Waters-Fuller, 2007, p.v). This *'slot'* ventilation is standard in UK double-glazed windows and complies with the minimum ventilation requirements under British Standards. The slot can usually be opened and closed by the user. A diagram of the windows tested can be found in the Appendix. The 6dB to an 11dB reduction in performance would suggest that a better tactic would be to separate the provision of light from the provision of natural ventilation and control the amount of noise entering through the natural ventilation opening. It may be possible to have a slot in the window on the quiet side and forgo a slot in the window on the noisy side, but it really depends on the specific project context and design. Given that the actual reduction varies so widely from the assumption, it would be prudent to design more robust responses to account for any variability.

There is a real requirement and standard for ventilation of rooms that slot vents fulfil. Still, there are many ways to meet a need, and looking to history and other jurisdictions may help illuminate other solutions. The inventiveness of humans, combined with the probability that someone in the past has been confronted with similar challenges before, means that the solutions probably already exist. Sometimes in solving one problem, we inadvertently create another, or in dispensing with an old technology for a new one, we lose a certain benefit. It isn't until much time has passed that we realise there was great wisdom in old ways of doing things, and that we resurrect some practices. In the next section, I will look to history and solutions from elsewhere in the world and provide insight into older solutions and some very modern innovations in window design.

5.7.2 DOUBLE-FRAME WINDOWS

Having two window frames, one on the outside and one on the inside, with an airspace between of approximately 100 to 200 mm are called by different names and can be variously referred to as double-

FIGURE 44 – TRADITIONAL DOUBLE-FRAME WINDOWS WITH 'PORTHOLE' OPENING WINDOW SECTION IN PRAGUE;PHOTO BY J.T. COOKE; MAY 2007 © J.T. COOKE. *NOTE: THE DOUBLE-FRAME WINDOWS WERE COMMON THROUGHOUT NORTHERN EUROPE. MANY OF THE DOUBLE-FRAMED WINDOWS HAD A SMALL PORTHOLE WINDOW THAT OPENED OUTWARD AND IS WITHIN THE LARGER OUTWARD OPENING WINDOW SECTION. THESE ARE CALLED TUULETUSIKKUNA (VENT WINDOWS) IN FINLAND, AND FORTOCHKA IN RUSSIA. THE POSITION OF THE 'PORTHOLE' WINDOW VARIES, AND IN THESE EXAMPLES, IN PRAGUE, THEY ARE IN THE BOTTOM LEFT PANE, BUT SOMETIMES THEY ARE IN THE TOP-RIGHT PANE.*

windows, double-frame windows, secondary windows, secondary glazing and coupled windows. In this guide they will be referred to as double-frame windows.

Figure 45 - A traditional double-frame window in an old building (circa 1825-1860) Porvoo (Borgå) Finland from the time Finland was a Grand Duchy of Russia; by J.T. Cooke; July 2018 © J.T. Cooke. *Note: Carl Ludvig Engel designed the master plan for the new-town area of Porvoo (Borgå) on a grid pattern in 1832, and at that time, Finland was a Grand Duchy in the Russian Empire. The windows may also date from this period. These windows are double-frame windows but do not include the smaller 'porthole' window for ventilation.*

Double-frame windows can be seen as a precursor to double glazed windows and are common in regions with very cold winters, such as; the alps, the lands of the former Hapsburg Empire, Germany, Scandinavia, Russia and the author has seen this type of window from the former Yugoslavia to Finland, and Russia to France, see Figure 44 to Figure 46. In considering how the double-frame window came about it may be that the traditional external shutters seen in southern Europe were the basis for the double-frame window and that the timber louvres in shutters were replaced with glass panes to create double-frame windows, see Figure 41, Figure 47, Figure 48. Moving from Italy and southern Europe to the colder climates of the Alps and northern Europe, it is not difficult to see how the external shutters of southern Europe were replaced with glazing and evolved to form a second set of windows, see Figure 49 to Figure 51.

There has been a considerable amount of research into one variation of the double-frame window that has two smaller opening windows within each double-frame window and can be seen in Figure 51. In Russia, they are called Fortochka, and in Finland, they are called Tuuletusikkuna (Too-le-toos-ik-oona), which translates as airing, or ventilation window, see Figure 46. The Russian name for the Fortochka windows may have its origin in the German word Pförtchen, meaning a portal or opening, and may have a seafaring meaning as a port-hole window in the door of a ship (Hilpert, 1845), see Figure 52. Today, the terms plenum-window, or acoustic-window are being used for this type of window, but they are a type of double-frame window and could be called a vent port-hole window.

Figure 46 - A traditional tuuletusikkuna double-frame window in an old building (circa 1825-1860) Porvoo (Borgå) Finland from the time Finland was a Grand Duchy of Russia, by J.T. Cooke; July 2018 © J.T. Cooke. *Note - the multiple opening sections. The individual window-pane in the outside window frame opens outward, and the lowest pane of the internal window frame can also be opened and a corresponding opening inside, see Figure 51. The building and window may date from the time Finland was a Grand Duchy in the Russian Empire, and the window design is similar to those found in Russia, the Czech Republic & Austria.*

The author has seen these windows in Austria, and the Czech Republic, see Figure 44, and there is a reference to them from 1839 in Vienna (Wilkey, 1839). The tuuletusikkuna window may be an evolution of traditional timber external window shutters where glass has replaced timber louvres. The traditional tuuletusikkuna, or fortochka double-frame window not only has good noise reduction characteristics while open and providing natural ventilation, the space between the two windows can act as a heat exchanger, warming or cooling the incoming air.

FIGURE 47 - TRADITIONAL SHUTTERED WINDOWS IN PORTOFINO, ITALY; PHOTO BY J.T. COOKE; AUG 2016 © J.T. COOKE. *NOTE:THE TRADITIONAL WINDOW AND SHUTTER ARRANGEMENT IN SOUTHERN EUROPE. THE SHUTTERS IN THIS PHOTO ARE NEW ALUMINIUM VERSIONS OF THE TRADITIONAL DESIGN. THE INNER WINDOWS ARE FULLY OPEN TO LET AIR IN. THERE ARE TWO FRAMES: AN OUTER SHUTTER FRAME AND AN INNER WINDOW FRAME. THE LOWER PANEL OF THE LOUVRES TILTS OUTWARD TO PROVIDE VENTILATION WITHOUT HEAT GAIN; THE 'PORTHOLE' IN THE TUULETUSIKKUNA WINDOW COULD BE A GLAZED EVOLUTION OF THIS ARRANGEMENT.*

Figure 48 - traditional shuttered windows in Portofino, Italy; Photo by J.T. Cooke; Aug 2016 © J.T. Cooke. *Note: The traditional window and shutter arrangement in southern Europe provides natural ventilation while shading the internal space. Occupants can open the shutters fully or have them closed and tilt the louvres for privacy or daylight access. Occupants can also tilt the lower louvre panel outward to increase the air intake while providing solar shading, as illustrated in the bottom window. It isn't difficult to see how the lower top-hung outward opening shutter became another glazed window in colder climates but was changed from the top-hung to side hung because there was less need for solar shade and ventilation. The creators of the tuuletusikkuna window replaced the louvre blades in the shutters with glass to provide greater insulation from the cold.*

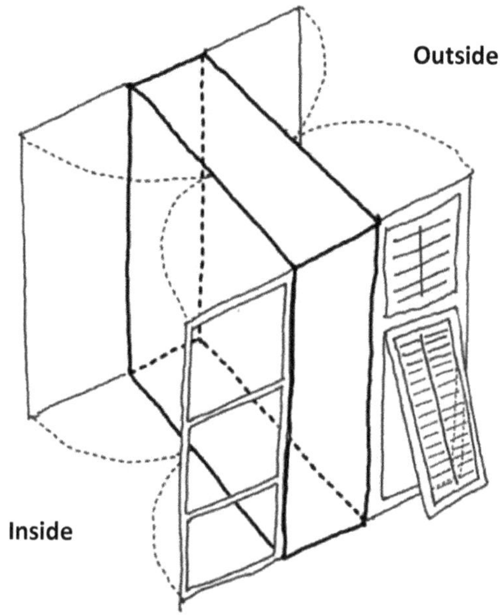

FIGURE 49 - EVOLUTION OF THE TUULETUSIKKUNA WINDOW; BY J.T. COOKE; JULY 2018. *NOTE: A TRADITIONAL INWARD OPENING WINDOW WITH OUTWARD OPENING SHUTTERS FOUND THROUGHOUT SOUTHERN EUROPE.*

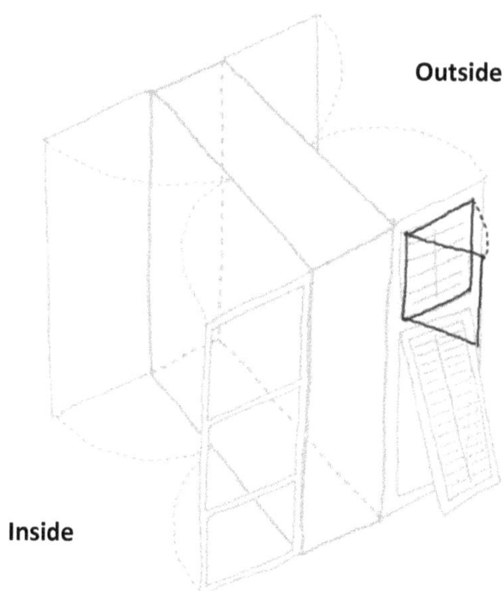

FIGURE 50 - EVOLUTION OF THE TUULETUSIKKUNA WINDOW. BY J.T. COOKE; JULY 2018. *NOTE: THE TRADITIONAL WINDOWS IN SOUTHERN EUROPE WITH SHUTTERS COULD HAVE EVOLVED INTO TUULETUSIKKUNA, AND THE TIMBER SHUTTERS COULD HAVE BEEN REPLACED WITH GLASS, WITH THE SMALLER VENTILATION WINDOW OPENING OUTWARD AS IN FIGURE 46 OR INWARD.*

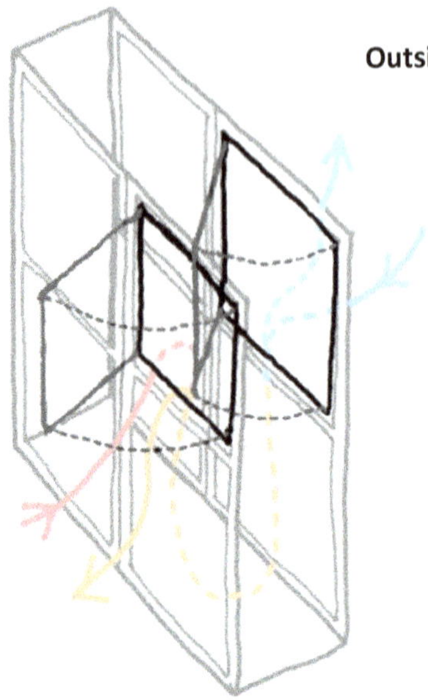

FIGURE 51 - EVOLUTION OF THE TUULETUSIKKUNA WINDOW; BY J.T. COOKE; JULY 2018. THE TRADITIONAL PORTHOLE WINDOW, CALLED TUULETUSIKKUNA IN FINNISH, OR FORTOCHKA IN RUSSIAN. *NOTE:IN COLD STILL (NO WIND) WEATHER, THE DOUBLE-FRAME WINDOW CAN ACT AS A HEAT EXCHANGER, AND THE OUTGOING, WARM AIR CAN PARTLY HEAT THE INCOMING FRESH AIR. HOWEVER, THIS DOES NOT WORK AS WELL IN WINDY CLIMATES. THERE HAS BEEN MUCH RESEARCH INTO USING A SIMILAR ARRANGEMENT OF WINDOWS AS A NOISE ABSORPTION CHAMBER, REDUCING NOISE INGRESS FROM OUTSIDE.*

The traditional use of double-framed windows in Scandinavia[1] (see Figure 45 and Figure 46) has stimulated the development of using the space between them as a sound absorption chamber, as illustrated in Figure 53 to Figure 57. The use of the space between window frames for noise reduction is a fascinating tactic, and the author has childhood memories of living in London in a house near a busy road, with a similar arrangement using double-framed windows called secondary double-glazing. Much research into augmenting window

---

[1] As mentioned previously, these types of windows were also common in the former territories of the Hapsburg Empire, Germany, Russia and a good deal of central Europe and helped insulate from the cold winters.

design including retrofitting existing windows with absorptive material has been done in Denmark and Germany. For further information about some Danish research see Søndergaard and Egedal (2016, pp. 1173-1183) and Rasmussen (2015, pp. 19-55), and see Figure 53 to Figure 57.

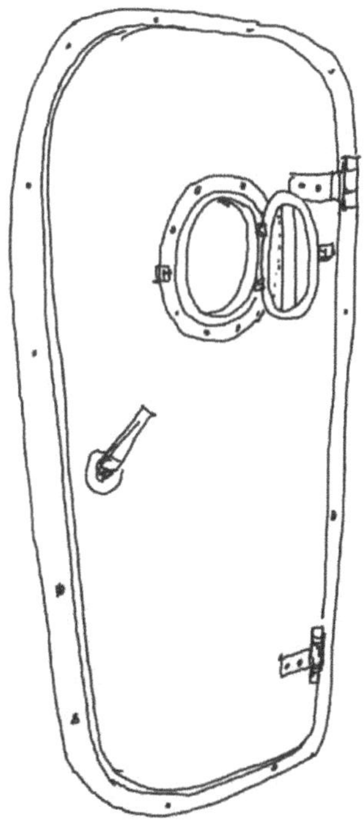

FIGURE 52 - A TYPICAL SHIPS DOOR WITH A PORTHOLE WINDOW IN IT; BY J.T. COOKE; JULY 2018 © J.T. COOKE. *NOTE: THE GERMAN WORD PFÖRTCHEN WAS USED TO REFER TO SHIPS PORTHOLES IN THE PAST (HILPERT, 1845) AND MAY HAVE BEEN THE ORIGINS OF THE RUSSIAN WORD FOR DOUBLE-FRAME WINDOWS WITH A PORTHOLE IN THEM, CALLED TUULETUSIKKUNA IN FINNISH AND FORTOCHKA IN RUSSIAN.*

Recalling the section *Some Principles Shapes*, the double-framed windows with absorption between is the snorkel type of absorption chamber in operation. The variability in the length of the absorption chamber is useful as it allows calibration of the window configurations to suit the context and noise reduction required. A more elongated chamber is possible for more significant noise reduction and longer sound waves, as illustrated in Figure 30, Figure 31 & Figure 55. Some examples of the use of double-frame windows in Denmark include Bikuben College hall of residence and Kollektivhuset in Copenhagen.

Figure 53 - Innovative Double-frame window in Bikuben College, Ørestad Copenhagen using the space between the glazing as a noise absorption chamber; image courtesy of Kelley Shaw ©; 2011. *Note: the open window can foul (hit) the blinds, and higher air flow can move the blinds compromising the ability to block out light, or alternately the blinds can reduce the free flow of air. This window arrangement is similar to the traditional tuuletusikkuna window see Figure 46.*

The Bikuben College halls of residence are located on the junction of two busy and noisy roads: Amagerfælledvej and Njalsgade. The building completed in 2006 was design by AArt Architects, and the design makes use of the thickness of the façade to incorporate double-framed windows. The space between the windows is used as a sound absorption chamber and allows natural ventilation and acoustic privacy from the traffic noise. The building is in a courtyard form, and the internal courtyard provides a central focus for residents.

Figure 54 – The courtyard of Bikuben College, Ørestad Copenhagen; image courtesy of Kelley Shaw ©; 2011. *Note: the Courtyard provides a focus for residents.*

Images of the innovative window design can be seen in Figure 53 and Figure 55, and note the snorkel shape path the air takes through the sound absorption chamber to absorb sound energy and provide air. The context of Bikuben College is close to the centre of Copenhagen, and separation from the noise source wouldn't be a viable option. In this settlement centre context, double-frame windows are an excellent option to achieve natural ventilation and noise reduction concurrently. As noted previously, the distance between the two opening windows could potentially be increased or decreased to calibrate the chamber for the wavelengths of sound being absorbed. The window design is brilliant, and the ability to tailor the window positions for different sound waves and frequencies makes it adaptable for use in different acoustic contexts. Designers could potentially use this type of solution in buildings with a double layer of curtain wall glazing.

Readers can inspect the context of the project at the following coordinates:

55°39'46.5"N 12°35'37.3"E

and the following address:

Bikuben Kollegium, Amagerfælledvej 50A, 2300 København, Denmark

and the following link:

Google.com/maps

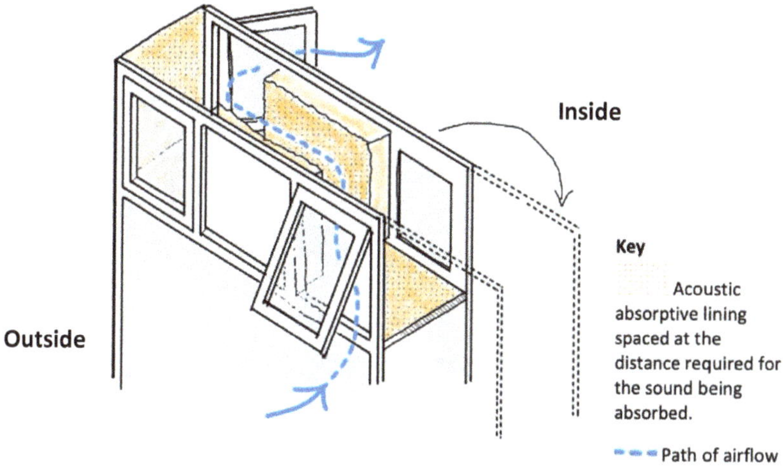

FIGURE 55 - DIAGRAM OF INNOVATIVE DOUBLE-FRAME WINDOW IN BIKUBEN COLLEGE, ØRESTAD COPENHAGEN USING THE SPACE BETWEEN THE GLAZING AS A NOISE ABSORPTION CHAMBER, ADAPTED FROM 'SOUND INSULATION OF THE BUILDING ENVELOPE SBI INSTRUCTION NO. 244'(P107) 1ST ED., BY B. RASMUSSEN & C. MØLLER PETERSEN. ORIGINAL IMAGE © SBI PUBLISHING 2013. REPRODUCED WITH THE PERMISSION OF SBI. HTTPS://SBI.DK/ANVISNINGER/PAGES/244-LYDISOLERING-AF-KLIMASKAERMEN-1.ASPX#/APPENDIKS-D-VINDUERS-LYDISOLATION *NOTE: THE OPEN WINDOW CAN FOUL (HIT) THE BLINDS AND HIGHER AIR FLOW CAN MOVE THE BLINDS COMPROMISING THE ABILITY TO BLOCK OUT LIGHT, OR ALTERNATELY THE BLINDS CAN REDUCE THE FREE FLOW OF AIR. THIS WINDOW ARRANGEMENT IS SIMILAR TO THE TRADITIONAL TUULETUSIKKUNA WINDOW SEE FIGURE 46.*

The second example is the refurbishment of an existing, late 1950's building, a Collective House (Kollektivhuset). The Kollektivhuset is located adjacent to the busy motorway 19, Lyngbyvej, and the mainline train station of Ryparken, which create a real challenge in providing natural ventilation and reduced noise. Because the project was a refurbishment rather than a new build, there was the additional challenge of having to retrofit any acoustic attenuation device. The window can be seen in Figure 56 and Figure 57.

The solution in this instance was a double-frame window with a pop-out window box to create a sound absorption chamber. Because the project was a refurbishment, a pop-out window gave sufficient space to create a sound absorption chamber. The window configuration allows for natural ventilation with reduced noise while the windows are open. The window configuration also allows the larger window section to open to allow cleaning and purge ventilation. Unfortunately, as with several of the double-frame window solutions, the fouling of blinds or curtains in the open ventilating position is an issue, but designers could overcome this with some minor changes to the design. Designers could also consider the additional challenge of delivering night-time cooling and darkness. This window solution is very impressive given the constraints of an existing building in an acoustically challenging context.

Readers can inspect the context of the project at the following coordinates:

55°42'51.1"N 12°33'34.3"E

and the following address:

Hans Knudsens Pl. 1, 2100 København, Denmark

and the following link:

Google.com/maps

Figure 56 – Photo of Double-frame window in the Kollektivhuset Hans Knudsens Plads 1, Copenhagen using the space between the glazing as a noise absorption chamber, image by L. S. Søndergaard © of DELTA Technology, 2014, Copenhagen: Reproduced with the permission of Delta Technology & L. Søndergaard. Originally published in Russervinduer; Project No. 1609, 2014 (p.8), Edited by Søren Vase Legarth, 2014, Ministry of Environment and Food of Denmark; Environmental Protection Agency, retrieved from https://www2.mst.dk/Udgiv/publikationer/2014/11/978-87-93283-13-8.pdf *Note: This is an impressive solution given the constraints of an existing building and an acoustically challenging site. However, one issue with this arrangement is that the internal open window can foul (hit) the blinds, and higher airflow could move the blinds compromising the ability to block out light.*

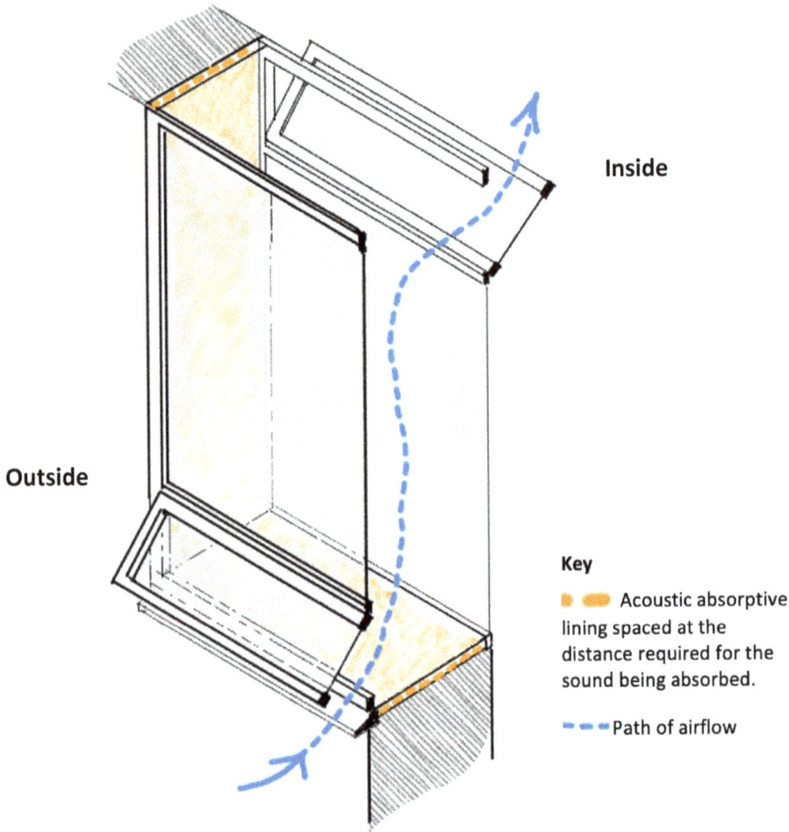

FIGURE 57 - DIAGRAM OF DOUBLE-FRAME WINDOW IN THE KOLLEKTIVHUSET HANS KNUDSENS PLADS 1, COPENHAGEN USING THE SPACE BETWEEN THE GLAZING AS A NOISE ABSORPTION CHAMBER, ADAPTED FROM SOUND INSULATION OF THE BUILDING ENVELOPE SBI INSTRUCTION NO. 244 (P107) 1ST ED., BY B. RASMUSSEN & C. MØLLER PETERSEN. ORIGINAL IMAGE © SBI PUBLISHING 2013. REPRODUCED WITH THE PERMISSION OF SBI. HTTPS://SBI.DK/ANVISNINGER/PAGES/244-LYDISOLERING-AF-KLIMASKAERMEN-1.ASPX#/APPENDIKS-D-VINDUERS-LYDISOLATION. *NOTE: ONE ISSUE WITH THIS ARRANGEMENT IS THAT THE OPEN WINDOW CAN FOUL (HIT) THE BLINDS AND HIGHER AIR FLOW CAN MOVE THE BLINDS COMPROMISING THE ABILITY TO BLOCK OUT LIGHT, OR ALTERNATELY THE BLINDS CAN REDUCE THE FREE FLOW OF AIR.*

In Germany, there have also been several innovative window designs using double-frame windows. Readers my recall the map of the HafenCity development from *section 3*. The HafenCity development in Hamburg spurred several new window designs, which allow natural ventilation and reduced noise levels with open windows. The

HafenCity Development is Europe's most extensive inner-city redevelopment. The development is located adjacent to the operational commercial port of Hamburg.

Readers can inspect the context of the project at the following coordinates:

53°32'27.3"N 9°59'17.9"E

and the following address:

Am Kaiserkai, 20457 Hamburg, Germany

and the following link:

openstreetmap.org

The night-time noise from the operations of the commercial port spurred the development of the HafenCity Window. The HafenCity development targeted 30dB internally in sleeping areas while occupants windows were open to provide natural ventilation. To enable the port's commercial activities to co-exist with residential development and provide natural ventilation at acceptable sound levels, the HafenCity Window was created by Eilenburger Fenstertechnik (window technology). The Eilenburger HafenCity Window is available in several configurations with varying sound reduction levels. The Eilenburger window configuration illustrated in Figure 59 is used in the HafenCity development. It claims to achieve a considerable 48dB noise reduction while the window is open, providing natural ventilation, depending on the noise source. The HafenCity window is another example of a double window-frame solution to reduce sound levels from inside to outside. One configuration of the HafenCity Window is similar to the Collective House window in Copenhagen and shares the snorkel shape path of airflow, illustrated in Figure 58. The Eilenburger Window is an excellent example of what can be achieved when strict regulation drives superior design and innovation.

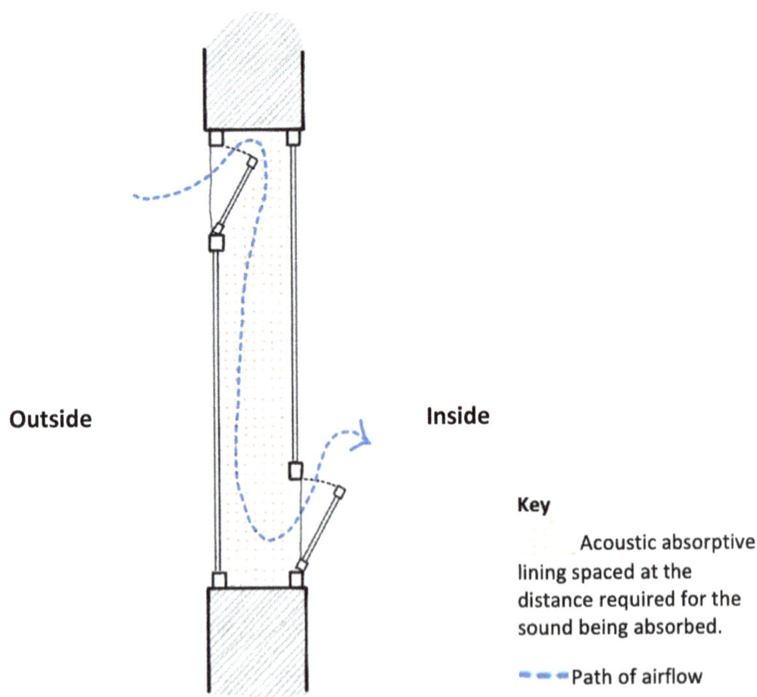

FIGURE 58 - A SECTION THROUGH A TYPICAL DOUBLE-WINDOW; BY J.T. COOKE; 2019 © J.T. COOKE; *NOTE: BOTH THE EILENBURGER HAFENCITY WINDOW AND THE COLLECTIVE HOUSE WINDOW IN COPENHAGEN SHARE THE 'SNORKEL' SHAPE OF THE AIR FLOW PATH TO ABSORB AS MUCH NOISE ENERGY AS POSSIBLE BEFORE THE AIR ENTERS THE INSIDE. ONE DIFFERENCE BEING THE OPENING DIRECTION OF THE WINDOWS.*

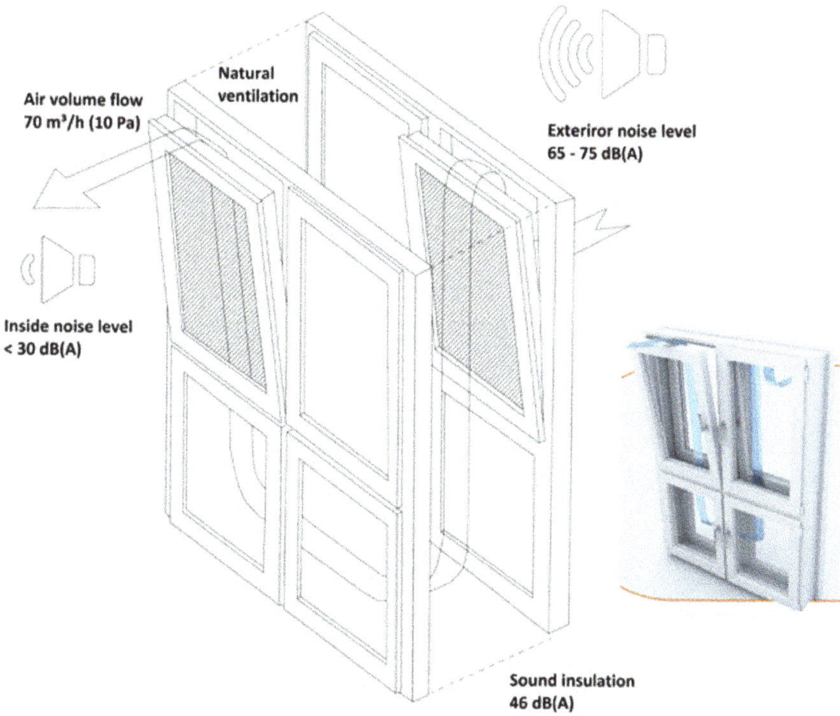

FIGURE 59 – NOISE ATTENUATING DOUBLE-FRAME WINDOW USED IN HAFENCITY DISTRICT REDEVELOPMENT IN HAMBURG, GERMANY. RESEARCH FOR THE DEVELOPMENT OF THE WINDOWS WAS PARTLY FUNDED FROM THE EU'S HORIZON 2020 RESEARCH AND INNOVATION AGREEMENT NO 783717. ADAPTED FROM SOUND-ABSORBING HAFENCITY WINDOWS FROM EILENBURG (P1), BY EILENBURGER FENSTERTECHNIK, GERMANY, 2018, WWW.EILENBURGER-FENSTER.DE © EILENBURGER FENSTERTECHNIK. *NOTE: THE CIRCUITOUS ROUTE OF THE PATH OF AIR-FLOW TO ABSORB AS MUCH SOUND ENERGY AS POSSIBLE FROM THE INCOMING AIR. THE 'SNORKEL' SHAPE IS ALSO EVIDENT IN THIS DESIGN. THIS WINDOW ARRANGEMENT IS SIMILAR TO THE TRADITIONAL TUULETUSIKKUNA WINDOW SEE FIGURE 46.*

In the previous section, *Some Principles Shapes* we discussed the snorkel shape for sound absorption. The Danish and German double-frame windows all make the air-flow in a snorkel shape through a sound absorption chamber to absorb as much sound energy as possible. The Kollektivhuset and HafenCity windows are similar, and readers may notice that the Bikuben College window is like one of the Kollektivhuset, or HafenCity windows laid horizontal to create air-flow in a snorkel shape. Eilenburger Fenstertechnik produces a wide range of windows and makes a window configuration similar to the Bikuben College window.

FIGURE 60 - NOISE ATTENUATING DOUBLE-FRAME WINDOW, OR PIXEL WINDOW USED IN AN APARTMENT BUILDING DEVELOPMENT AT 17-25 BARMBEKER STRASSE, IN HAMBURG, GERMANY. PHOTO BY LARS ROHDE, PROJECT BY, AUMANN KATZSCH ARCHITEKTEN, PLANWERKEINS ARCHITEKTEN, WDK ARCHITECTS & ENGINEERS, STARCKE². © LARS ROHDE & AUMANN KATZSCH ARCHITEKTEN. *NOTE: IN THIS EXAMPLES THE 'POP-OUT' WINDOW ELEMENT IS THE ABSORPTION CHAMBER AND IS ALSO A DOUBLE FRAME WINDOW SOLUTION.*

The old double-frame window has been resurrected utilising the space between the window frames to absorb sound energy. Although this solution may not mitigate all sound ingress, it can considerably reduce noise exposure. The use of double-frame windows as sound absorption chambers is a practical solution where space for separation is constrained, as in city centres and in refurbishments of existing buildings where window opening positions may already be fixed.

Another innovative window design, also in Hamburg, can be seen in Figure 60 & Figure 61. The window was designed for an apartment development near a busy street in Barmbecker Strasse in Hamburg by Aumann Katzsch Architekten, Planwerkeins Architekten, WDK architects & engineers, & Starcke[2].

FIGURE 61 - NOISE ATTENUATING DOUBLE-FRAME WINDOW, OR PIXEL WINDOW USED IN AN APARTMENT BUILDING DEVELOPMENT AT 17-25 BARMBEKER STRASSE, IN HAMBURG, GERMANY. PHOTO BY LARS ROHDE, PROJECT BY, AUMANN KATZSCH ARCHITEKTEN, PLANWERKEINS ARCHITEKTEN, WDK ARCHITECTS & ENGINEERS, STARCKE[2]. © LARS ROHDE & AUMANN KATZSCH ARCHITEKTEN. *NOTE: THE NOISE ABSORPTION CHAMBER IS CREATED BY 'POP-OUT' WINDOW BOXES THAT CREATE A DYNAMIC MODULATION OF THE FAÇADE.*

The design also uses a double-frame window to create a noise absorption chamber that pops out from the main façade line. The small bottom hung window on the external window is the main path of air entry, and sound absorption occurs in the space between the window frames. The window is referred to as the 'pixel' window by its designers and is another excellent example of an innovative design using a double-frame window. The pixel window and the Kollektivhuset window use a pop-out window box to create a sound absorption chamber. This is a useful technique for building refurbishments or where space for a chamber is constrained.

Readers can inspect the context of the project at the following coordinates:

53°32'27.3"N 9°59'17.9"E

and the following address:

Am Kaiserkai, 20457 Hamburg, Germany

and the following link:

openstreetmap.org

In Hong Kong, there have also been window innovations using double-frame windows that make air-flow in a snorkel shape. Hong Kong is the quintessential high-density urban centre, and the high density has spurred some very innovative solutions to noise using double-frame windows. The Hong Kong Housing Authority, Environmental

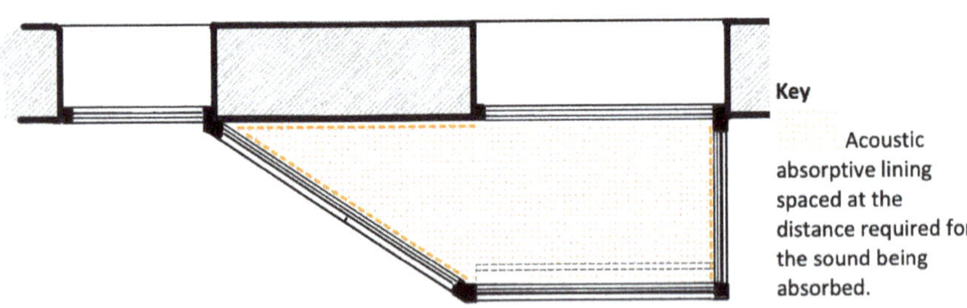

FIGURE 62 – A DIAGRAMMATIC PLAN OF THE WINDOW NOISE ATTENUATING DOUBLE-FRAME WINDOW, OR PIXEL WINDOW USED IN AN APARTMENT BUILDING DEVELOPMENT AT 17-25 BARMBEKER STRASSE, IN HAMBURG, GERMANY; IMAGE BY J. T. COOKE; NOV 2020 © J.T. COOKE. *NOTE: THE WINDOW BOX IS LINED WITH SOUND ABSORPTIVE MATERIAL. THERE IS AN INWARD TILTING WINDOW AT THE BOTTOM OF THE EXTERNAL BOX WINDOW, AND AN UPPER OPENING SECTION AT THE TOP OF THE BOX WINDOW. THE LARGER PIVOT SECTION CAN BE FOR CLEANING AND PURGE VENTILATION.*

Protection Department and others have researched creative window arrangements, see (HL HO & WM CHU, 2016) & (Yeung, 2016) as examples. Figure 63 illustrates one window investigated that uses double-frame windows. The window uses a similar tactic to outer double-frame windows in utilising the space between the windows as a sound absorption chamber. The window has a simple sliding section internally and an outward opening section externally, but rain ingress may be an issue.

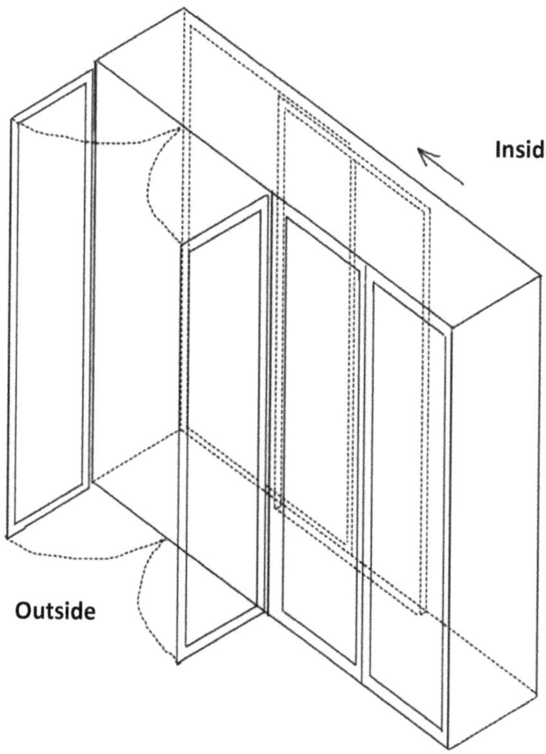

FIGURE 63 – DIAGRAM OF A DOUBLE-FRAME WINDOW IN KING TAI COURT, HONG KONG; BY J.T. COOKE; BASED ON IMAGES COURTESY OF THE ENVIRONMENTAL PROTECTION DEPARTMENT AND THE HONG KONG HOUSING AUTHORITY (HKHA), HKSAR GOVERNMENT; FROM HTTPS://WWW.EPD.GOV.HK/EPD/INNOVATIVE/GREENY/ENG/CONTENT/KING-TAI-COURT.HTML?TYPE=FTAB_20 © 2016 ENVIRONMENTAL PROTECTION DEPARTMENT. *NOTE: ONE ISSUE WITH OUTWARD OPENING OUTER WINDOWS MAY BE RAIN INGRESS INTO THE SPACE BETWEEN WINDOW FRAMES. A POTENTIAL 4-8DB(A) NOISE REDUCTION IS NOTED ON THE ENVIRONMENTAL PROTECTION DEPARTMENT WEBSITE.*

Another innovative window design is the Hush Window™. The Hush Window™ also uses a double-framed window and includes an innovative absorption chamber that can be tailored to suit the specific sound context. The Hush Window™ comes in a variety of arrangements to suit new-build and refurbishment developments. The window includes features to prevent disturbance of curtains or blinds by the air movement and can be adapted to the acoustic context by the user over time.

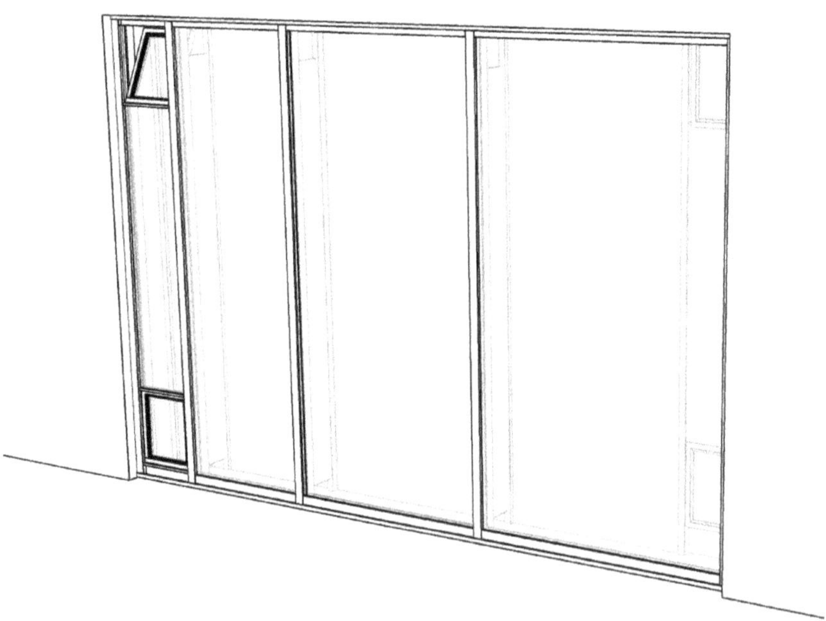

FIGURE 64 – A DIAGRAM OF A LARGER HUSH WINDOW™; COURTESY OF HUSH WINDOWS; JULY 2019 © HUSH WINDOWS. *NOTE: THIS IS AN ANOTHER EXAMPLE OF A DOUBLE-FRAME WINDOW. THE STAGGERED ARRANGEMENT OF THE VENTILATION WINDOWS CREATES A SNORKEL AIR-FLOW PATH FOR THE INCOMING AIR, WHICH ABSORBS THE SOUND ENERGY FROM THE INCOMING AIR.*

The Hush Window ™ also seeks to use the space between the window frames as a heat exchange, allowing the incoming fresh air to be pre-heated or cooled by the outgoing air, taking inspiration from the old Tuuletusikkuna window. Further information is available on the Hush Windows ™ website:

hushwindows.com

Figure 65 - A diagram of a smaller Hush Window™; courtesy of Hush windows; July 2019 © Hush Windows. *Note: This is an another example of a double-frame window. The staggered arrangement of the ventilation windows creates a snorkel air-flow path for the incoming air, which absorbs the sound energy from the incoming air.*

All of these types of double-frame windows can be an excellent option in refurbishments where there are existing openings in walls, and making new openings is difficult. These windows are also useful in new builds, particularly where there is limited external façade surface for openings. The increase in the thickness of exterior walls because of more significant insulation requirements affords an opportunity to use the thickness of the wall as a sound absorption chamber with double-frame windows.

Architects and designers may need just one or two acoustically attenuated windows at the intake and possibly exhaust points. Only the designated ventilation openings need to be acoustically attenuated windows. Designers could position openings to maximise the natural air movement process they are using while minimising noise exposure; refer to the section on *natural ventilation*. As noted in the section on natural ventilation, designers need to carefully choose the intake and exhaust points for natural ventilation and consider the natural driver of air movement on which they are relying.

Designers may choose to combine a double-frame window with an acoustically attenuated passive vent as an alternative to two double-frame windows. As illustrated in Figure 15, the designers may be using buoyancy to move air and choose to locate the exhaust vent in a roof. Combining a double-frame window with an acoustically attenuated passive vent could be more appropriate in areas such as a roof and may reduce overall cost. I will discuss acoustically attenuated passive vents in the next section of the book.

5.7.3 ACOUSTICALLY ATTENUATED PASSIVE VENTILATORS

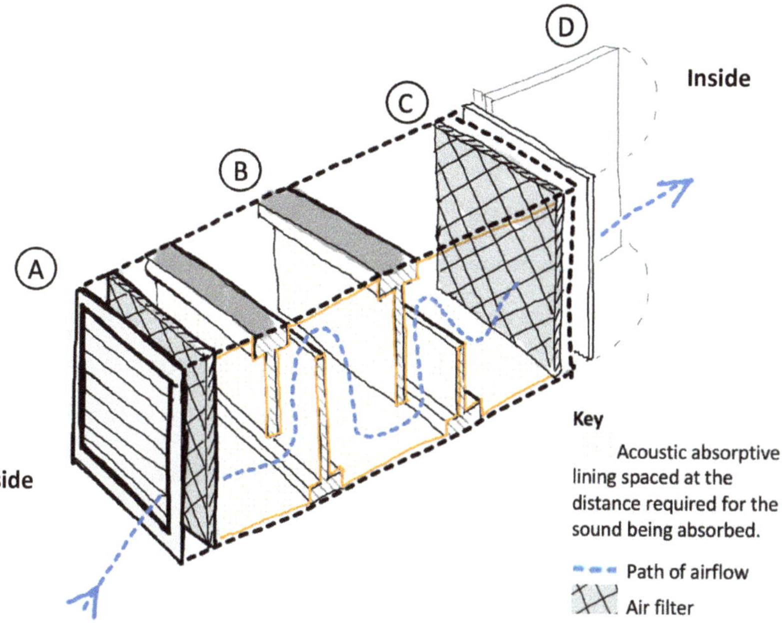

FIGURE 66 - A TYPICAL ACOUSTICALLY ATTENUATED VENT WITH AN OUTER AND INNER LOUVRE GRILLE AND SOUND ABSORBING ATTENUATION LAYERS BETWEEN; BY J.T.COOKE; JUNE 2019 © J.T. COOKE.
NOTE: THE EXTERNAL OPENING A IS USUALLY COVERED BY LOUVRES, INSECT MESH AND AN AIR FILTER. THE AIR PASSES THROUGH THE SOUND MAZE B AND THROUGH TO THE INSIDE. THE AREA SHOWN AS A SOUND MAZE COULD ALSO INCLUDE OR BE A SPLITTER LIKE THAT ILLUSTRATED IN FIGURE 70. THE INTERNAL OPENING C COULD HAVE AN INSECT MESH, AIR FILTER AND A THERMALLY SEALABLE HATCH (DOOR) D TO KEEP HEAT, OR COOL INSIDE.

Basic Principles

- **Position** - Can be at 90 degrees to noise source to improve performance.
- **Indirect Air Path** - Create an absorption maze to reduce the amount of sound energy getting through.
- **Air Flow Rate** – One issue is that ventilators can have limited airflow and may require some mechanical ventilation assistance to achieve flow rates.

Another way of providing ventilation and acoustic privacy is with the use of acoustically attenuated ventilators. Acoustically attenuated vents have several names and are sometimes called silencers, or mufflers. Acoustically attenuated passive ventilators are easily included in new builds because the designer can include and locate additional openings to suit the requirements of passive ventilators. They can be included in refurbishments; however, in many refurbishments the window openings already exist and a continued requirement for daylight and ventilation favours the use of double-frame windows. As mentioned previously, one option may be to include an acoustically attenuated passive ventilator in the roof to utilise any convection currents to drive air movement.

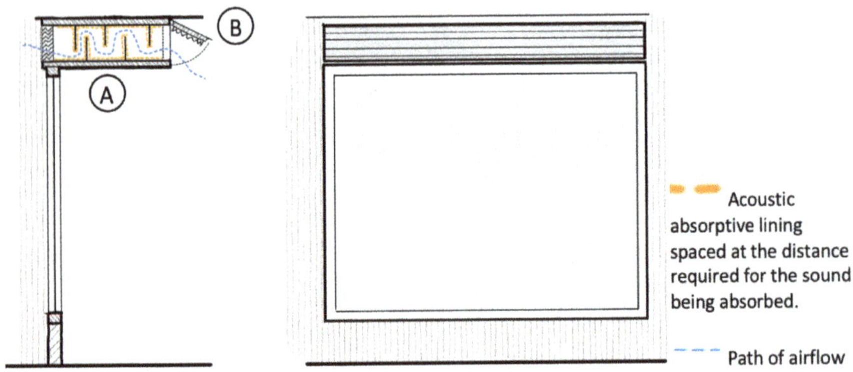

FIGURE 67 – SECTION AND ELEVATION OF AN ATTENUATED VENTILATOR ABOVE A WINDOW; BY J.T. COOKE; NOV 2019 © J.T. COOKE. *NOTE: THE LOCATION OF THE VENTILATOR WILL DEPEND ON THE FUNCTION OF THE OPENING AS AN AIR INTAKE OR EXHAUST, AND THE FORCE BEING USED TO MOVE THE AIR. A THE ACOUSTICALLY ATTENUATED VENTILATOR BOX. B THE OPENING HATCH, WHICH MUST BE CAPABLE OF PROVIDING AN AIR-TIGHT AND THERMAL SEAL.*

Acoustically attenuated vents are used in passive ventilation systems that rely on natural forces to move air, and active systems, such as mechanical ventilation and HVAC systems, which depend on electricity to drive mechanical devices to move air. In this section, we will primarily look at passive systems, but with the addition of hybrid systems that rely on mechanical fans already in use in bathrooms and kitchens. Hybrid systems can be helpful because relying on natural forces alone can be challenging to move air predictably. One hybrid system is to make use of the pressure difference created by extract fans that are regularly installed in bathrooms and kitchens. The extract fans in kitchens and bathrooms are usually installed to remove air pollutants, and odours, and create a negative pressure in kitchens and bathrooms to ensure that air passes into these rooms rather than out of them. Hybrid systems make use of the negative pressure already created by extract fans to draw air in from outside through passive acoustic vents. The sustainability argument of hybrid systems is that the bathroom and kitchen exhaust fans will be using electricity anyway.

Passive acoustically attenuated vents primarily rely on a pressure difference to move air, and the pressure difference can be caused by several natural phenomena, such as:

- by wind creating a pressure differential
- by solar radiation causing one side of a building to be warmer than the other creating a differential.
- Caused by heating the internal space creating a pressure differential between inside and outside.
- Caused by heat from people and appliances that create an air buoyancy effect inside a room.

The section on *natural ventilation* discussed the principles of natural ventilation, and Figure 15 illustrated some of the drivers of natural ventilation. The natural phenomena that drive air in a passive system can be challenging to predict and, consequently, can be less convenient or immediate than mechanical systems; however, passive systems are a considerably cheaper and low maintenance option than HVAC. As discussed in the section on *natural ventilation*, when choosing a location for a vent, it is essential to understand if the vent is an intake, exhaust, or both because it will influence the position of the vent.

In the section on *Simple Techniques for Locating Ventilation Openings and Windows,* we discussed how the thoughtful location of ventilation openings could significantly reduce the noise reaching the opening, and the same is true when locating ventilators.

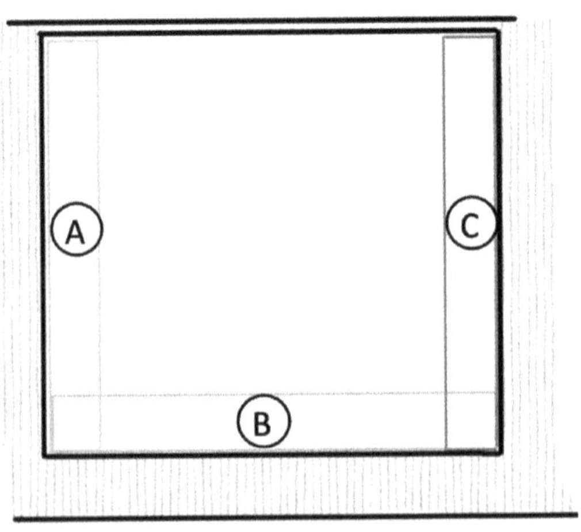

FIGURE 68 – AN ELEVATION OF THE VARIOUS POSITIONS FOR ATTENUATED VENTILATORS; BY J.T. COOKE; NOV 2019 © J.T. COOKE. *NOTE: THE LOCATION OF THE VENTILATOR WILL DEPEND ON THE FUNCTION OF THE OPENING AS AN AIR INTAKE OR EXHAUST, AND THE FORCE BEING USED TO MOVE THE AIR. THE LOCATION OF THE VENT WILL ALSO DEPEND ON THE PARTICULAR PRACTICAL REQUIREMENT OF THE CONTEXT. A & C ARE SIDE VENTILATORS, B IS A BOTTOM VENTILATOR.*

As illustrated in Figure 67 and Figure 68 ventilators can be located above, or below, or to the side of windows, or as shown in Figure 69 that can be separated from the window and located at 90 degrees to the noise source to reduce the noise reaching the opening. Designers may not have to make every window or vent an acoustically attenuated opening. It may be possible to choose specific openings to be acoustically attenuated, which could be just one opening per room. Designers may only need one or two for air intake and exhaust in open plan living areas in apartments. By carefully locating openings to suit the natural ventilation requirements and while considering the noise source, designers can significantly reduce the amount of sound energy before the acoustic vent has even absorbed any noise.

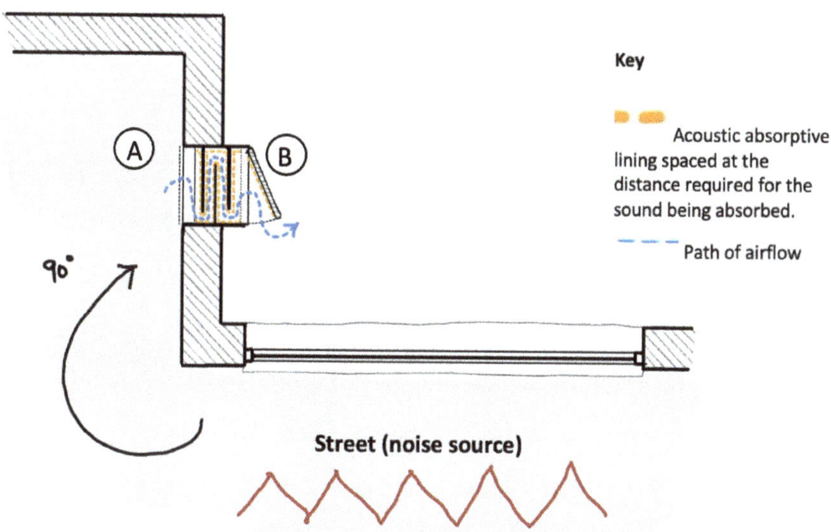

FIGURE 69 – A PLAN OF THE LOCATION OF AN ATTENUATED VENT AT 90 DEGREES TO THE SOURCE OF NOISE; BY J.T. COOKE; NOV 2019 © J.T. COOKE. *NOTE: THE ACOUSTICALLY ATTENUATED VENTILATOR CAN BE LOCATED AT 90 DEGREES TO A SOURCE OF NOISE TO REDUCE NOISE REACHING THE VENT AND SEPARATED FROM THE WINDOW, WHICH CAN BE SOLELY A SOURCE OF DAYLIGHT. A THE ACOUSTICALLY ATTENUATED VENT IS AT 90 DEGREES TO THE NOISE SOURCE, WITH WEATHERPROOFING AND INSECT MESH EXTERNALLY. B THE OPENING HATCH, WHICH MUST BE CAPABLE OF PROVIDING AN AIR-TIGHT AND THERMAL SEAL WHEN THE VENT IS NOT IN USE.*

Acoustically attenuated ventilators can be produced in many forms and sizes. As discussed in the previous section *Some Principles Shapes* most sound absorption happens in a few basic shapes and attenuators are generally no different. Attenuators can be as simple as acoustically treated louvres, or a sound absorption chamber, with a sound absorbing plenum, or sound maze (see Figure 31) inside the chamber, or a splitter type attenuator (see Figure 70), or a resonance absorber (see Figure 22 & Figure 26) as discussed in the section about *Absorption.*

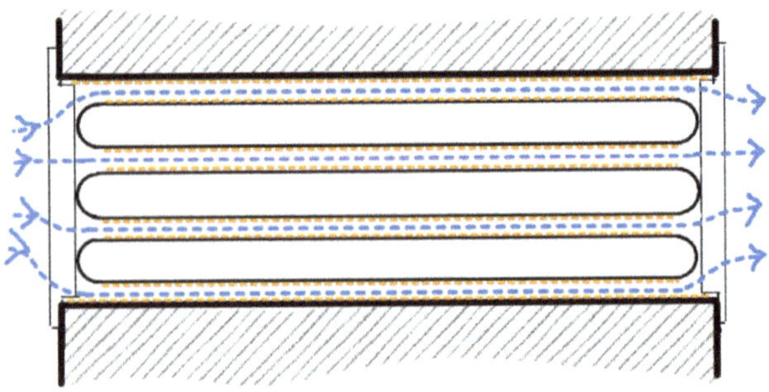

FIGURE 70 - A PLAN OF A TYPICAL SPLITTER; BY J.T. COOKE; NOV 2019 © J.T. COOKE. *NOTE: THE ABSORPTION OCCURS ALONG THE PATH THE AIR IS TRAVELLING. THE NARROW CHAMBERS BETWEEN THE FINS THE AIR HAS TO PASS THROUGH RESTRICT THE FREQUENCIES OF SOUND THAT CAN GET THROUGH THE NARROW CHAMBERS, THUS REDUCING THE SOUND INGRESS FROM OUTSIDE.*

Acoustically attenuated ventilators are not a new idea, and Figure 74 shows one in operation in the offices in Porvoo Finland (circa the 1970s). Figure 71 to Figure 73 illustrate some more recent examples are from the new Cambridge neighbourhood of Eddington. The Eddington Cambridge project includes some excellent innovations to provide natural ventilation while mitigating noise from the nearby M11 motorway. The Eddington Cambridge key worker housing designed by Mecanoo Architecten uses innovative acoustically attenuated vents set to the side of the window frame, see Figure 71 and Figure 72. The window side ventilator also includes a chiller to cool incoming air to tackle summertime overheating and provide night-time cooling. The separation of the source of ventilation from the source of light has benefits in that it allows ventilation and darkness simultaneously for night-cooling, see Figure 73. The window design also provides an aesthetically pleasing dynamic facade, see Figure 72.

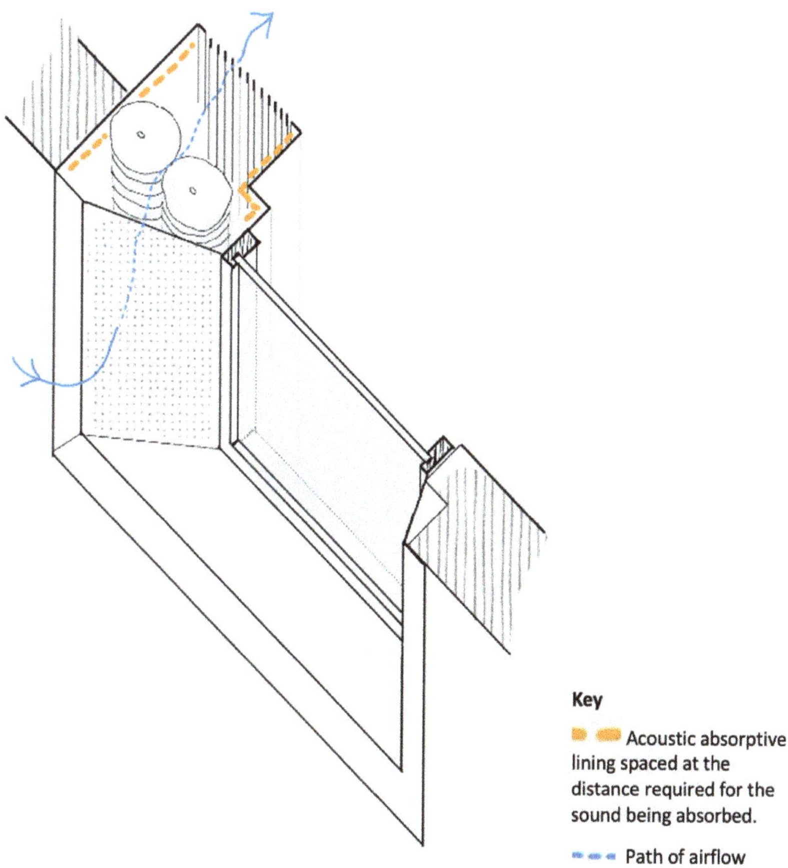

FIGURE 71 - SKETCH PLAN OF THE ACOUSTIC ATTENUATOR UNIVERSITY OF CAMBRIDGE; BY J. T. COOKE; BASED ON DRAWING COURTESY OF MECANOO ARCHITECTEN; MARCH 2018 © J.T.COOKE.

NOTE: THIS DESIGN HAS A SIDE ACOUSTICALLY ATTENUATED VENT THAT INCLUDES CHILLERS TO COOL THE AIR ENTERING AND IS PART OF THE NIGHT-COOLING STRATEGY. THE GLAZED WINDOW SECTION CAN ALSO BE OPENED FOR PURGE VENTILATION IF REQUIRED.

FIGURE 72 – ELEVATION OF KEY-WORKER HOUSING BUILDING IN THE EDDINGTON NEIGHBOURHOOD WITH NOISE ATTENUATING WINDOW REVEAL VENTS. REPRINTED COURTESY OF MECANOO ARCHITECTEN, BY MECANOO ARCHITECTEN, 2018, RETRIEVED FROM WWW.MECANOO.NL © MECANOO ARCHITECTEN.

FIGURE 73 - INTERNAL VIEW OF THE NOISE ATTENUATION VENT TO THE RIGHT OF THE WINDOW IN KEY-WORKER HOUSING IN CAMBRIDGE'S NEWEST NEIGHBOURHOOD; EDDINGTON, IMAGE COURTESY OF MEDIA FOR EDDINGTON DEVELOPMENT, UNIVERSITY OF CAMBRIDGE, 2018 © UNIVERSITY OF CAMBRIDGE. *NOTE: SEPARATING THE SOURCE OF VENTILATION FROM THE SOURCE OF LIGHT HAS BENEFITS IN THAT IT ALLOWS VENTILATION AND DARKNESS SIMULTANEOUSLY, WHICH IS PARTICULARLY USEFUL WHEN USING A 'NIGHT COOLING' STRATEGY.*

Readers can inspect the context of the Eddington, Cambridge project at the following coordinates:

52°13'00.5"N 0°05'09.9"E

and the following address:

Turing Way, Cambridge, UK,

and the following link:

openstreetmap.org

Figure 74 - Photo of acoustically attenuated ventilator in a window frame in civic offices in Porvoo (Borgå) Finland, by J.T. Cooke; July 2018 © J.T.Cooke. *Note: the ventilator at the bottom left of the window.*

5.8 BALCONIES

Using balconies as noise absorption chambers

Basic Principles

- **Absorption** - The balcony is a sound absorption chamber.
- **Indirect Air-Path** - The balcony becomes a sound absorption snorkel with the addition of another layer of glazing (see Figure 76).
- **Controlled Ventilation Route** - All natural-ventilation passes through the balcony, and the acoustic absorptive lining absorbs sound.

Balconies are ubiquitous with high-density dwellings and in particular in apartment living, and high-density dwellings are more likely in an urban centre where space is scarce. The use of the balcony as a sound absorption chamber is an innovative solution to mitigate noise using existing space in an apartment. The suitability and effectiveness of a balcony as an absorption chamber to mitigate noise ingress will depend on the level and frequency of the noise. Balcony absorption chambers are unlikely to be effective against deafening or intermittent loud noise, but they can be helpful in some contexts. The use of a balcony as a sound absorption chamber combined with an acoustically attenuated passive vent, as illustrated in Figure 76, may provide an improved solution where night-time ventilation is required because it allows darkness and ventilation concurrently.

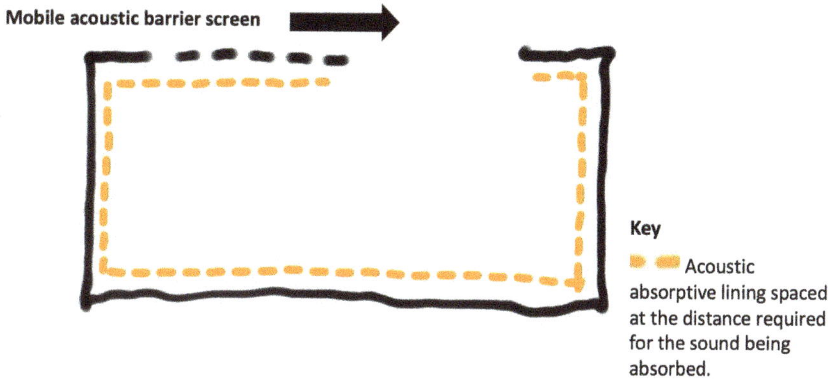

Figure 75 - Balconies used as an absorption chamber, by J.T. Cooke, a typical building situation with windows facing the noise source; June 2018 © J.T. Cooke. *Note: a 'wintergarden' with a screen can allow users to alter the level of acoustic privacy or ventilation, however people cannot predict when sudden noises beyond their control happen, and so the ability to make pre-emptive adaption for acoustic privacy is questionable. The effectiveness of the mobile screen is also determined by its air-tightness because sound waves move through the air, and if the air can enter then so can sound.*

Sound absorptive balconies take several forms, and Figure 75 illustrates a diagram of a generic sound absorptive balcony. Generally, acoustic absorbent balconies are lined with sound absorbent material to absorb sound energy as the air passes through the balcony. Some have an outer mobile acoustic-barrier screen that faces the noise source as illustrated in Figure 77. The effectiveness of the mobile acoustic-barrier screen to form a barrier to sound energy is reliant on the airtightness of the screen. Dependant on the level and frequency of noise to be reduced, some balconies only require an acoustic absorptive lining, as illustrated in Figure 81. Using balconies as an absorption chamber or plenum is a similar concept to a double-frame window, just more extensive, with the enclosure created by the external wall of the building and the balustrade of the balcony doing the same job as the two window frames in the double-framed window. There have been some fascinating, innovative solutions in the EU,

Hong Kong, and Australia. In 2016 the EU funded SONORUS Project researched several balcony balustrade profiles designed to reflect sound. As a part of their research, the SONORUS Project investigated several balcony balustrade profiles to assess their effectiveness at reflecting noise emanating from the traffic below. A similar study in Hong Kong in 2016 used the shape of the balustrade to reflect noise down toward the motorway traffic and used the balcony as a sound absorption chamber, as illustrated in Figure 83; however, the noise mitigation was limited.

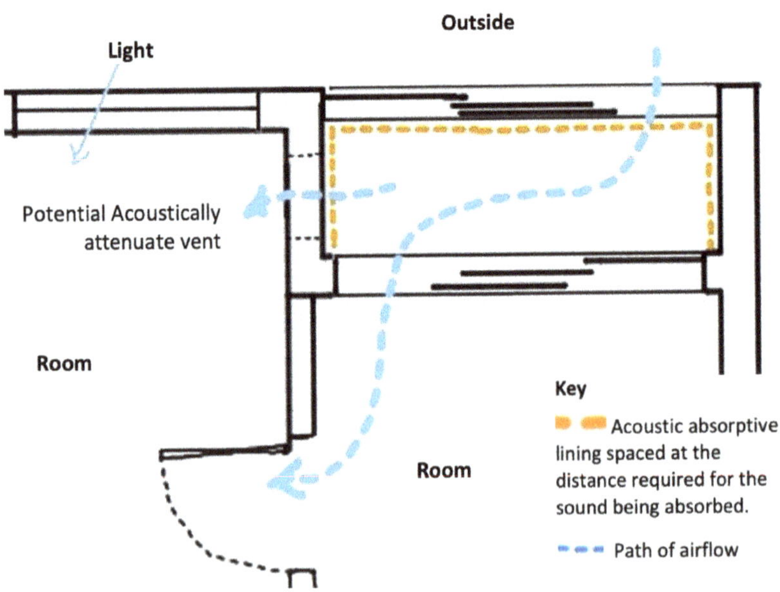

FIGURE 76 - BALCONY AS PLENUM ABSORPTION CHAMBER; BY J.T. COOKE; JUNE 2018 © J.T. COOKE. *NOTE: THE ARRANGEMENT OF SLIDING OUTER SCREEN WINDOWS AND INNER SLIDING DOORS FORM THE PATH OF AIR INTO A SNORKEL SHAPE AND USES THE BALCONY CHAMBER TO ABSORB NOISE. THE AIR FLOW IS CHANNELLED IN A SNORKEL SHAPED PATH TO MAXIMISE THE CONTACT WITH THE SOUND ABSORBING MATERIAL, AND REDUCE A DIRECT ROUTE FOR NOISE INGRESS. THE TACTIC HAS RECENTLY BECOME COMMON IN SYDNEY, AUSTRALIA.*

Figure 77 and Figure 78 illustrate two balconies in Sydney, Australia that use a mobile glazed area above the balustrade as an acoustic barrier, and both these examples operate in a similar manner to the sketch in Figure 76. As illustrated in Figure 76 the glazing on top of the balcony balustrade in addition to the glazing to the living area create a path of air flow in a snorkel shape forcing the air in an elongated route past the acoustic absorptive lining on the balcony walls and ceiling to absorb as much sound energy as possible. One issue with glazing above a balcony balustrade is that the balcony can become an enclosed room, which can be used unsightly if used as a storage area, as illustrated in Figure 78. Another issue with using balconies as a noise absorption chamber that faces the noise source is that residents' primary outdoor space will be subject to noise. The occupant must choose between being in an enclosed or semi-enclosed outdoor space, or subject to noise with the secondary glazing open. In warmer climates, this can lead to overheating when the screens are closed, and a better solution is thoughtful site planning and internal space planning to locate balconies on the quiet side, which I will address in *section 6.3*.

The inclusion of glazing to make balconies sealable is not a new idea and has been used in colder regions such as Scandinavia and Germany for many years to create protected outdoor space for the cold winters. Figure 79 and Figure 80 illustrate some examples of mobile glazing in balconies in use in Helsinki, Finland. The primary reason for mobile glazing in colder climates is to provide protection from the cold winters; however, they can also provide some acoustic attenuation. The glazing in Figure 79 is located near a noisy road and also provides some protection from noise. The effectiveness of the glazing to block sound waves depends on their airtightness, and if air can get through the glazing then so can sound waves.

FIGURE 77 - SLIDING GLASS SCREEN AS AN ACOUSTIC BARRIER IN A BALCONY; 222 ELIZABETH ST SYDNEY; BY PTI ARCHITECTURE, SYDNEY, PHOTO BY J.T. COOKE, JUNE 2019 © J.T. COOKE. *NOTE: THE BRICK BALUSTRADE HELPS TO REFLECT NOISE, AND THE MOBILE ACOUSTIC-BARRIER SCREEN ON TOP OF THE BRICK BALUSTRADE BLOCKS NOISE IN THE CLOSED POSITION AND REDUCES NOISE IN THE OPEN POSITION. THE INCLUSION OF A MOBILE EXTERNAL SUN-SHADE ALLOWS FOR NATURAL VENTILATION WHILE BLOCKING OUT SUNLIGHT. THE BALCONY CAN BECOME A SOUND ABSORPTION CHAMBER, BUT THE OCCUPANTS NEED TO UNDERSTAND HOW TO USE THE BALCONY FOR NOISE REDUCTION AND VENTILATION. SOME ISSUES WITH THIS TYPE OF SOLUTION CAN BE THAT THE OCCUPANT MUST CHOOSE BETWEEN BEING IN AN ENCLOSED OR SEMI-ENCLOSED OUTDOOR SPACE OR BEING SUBJECTED TO NOISE WITH THE GLAZED BARRIER IN THE OPEN POSITION.*

FIGURE 78 - MOBILE GLASS PANELS IN BALCONIES AS ACOUSTIC BARRIER IN DARLING SQUARE, SYDNEY, BY DENTON CORKER MARSHALL, SYDNEY; PHOTO BY J. T. COOKE JUNE 2019 © J.T. COOKE. *NOTE THE MOBILE GLAZED ACOUSTIC BARRIER ON TOP OF THE GLASS BALUSTRADE FOR THE FIRST TWO FLOORS IS INCLUDED TO REDUCE NOISE. THE APARTMENTS FACE A NEW PUBLIC SQUARE AND MAY HAVE BEEN SUBJECT TO SOCIAL NOISE FROM THE SQUARE AND RESTAURANTS BELOW. THE ENCLOSED BALCONIES CAN BECOME UNSIGHTLY STORAGE AREAS IF THEY ARE NOT USED AS BALCONIES. ONE ISSUE WITH THIS TYPE OF SOLUTION IS THAT THE OCCUPANTS' PRIMARY OUTDOOR SPACE FACES THE NOISE SOURCE.*

Figure 79 - Mobile glass panels in balconies at 5 Käskynhatijantie, Helsinki; Photo by J. T. Cooke; June 2018 © J.T. Cooke. *Note: The building is located in a fairly busy road and its context can be inspected in the address above. The mobile glass screens inside the glass balustrade that create a protected space and the inclusion of mobile external solar shades. The glazing is primarily provided for thermal comfort in winter.*

Figure 81 and Figure 82 illustrate an elegant solution to noise generated by a nearby tram stop. The underside of the balconies is treated with a sound absorbent material to create an outdoor balcony area that is less affected by noise. As illustrated in Figure 82, the inclusion of a laser-cut panel to hold the sound absorptive material makes for an elegant, and functional solution. Balconies of the type illustrated in Figure 81 and Figure 82 that do not require any user intervention can be practical solutions; however; as shown in Figure 76, balcony designs that rely on the occupant to move a mobile acoustic-barrier screen can be problematic. Acoustic-barrier screens require the occupant to move the barrier to create a larger or smaller opening to suit the natural ventilation or acoustic requirements and may overheat.

Figure 80 - Mobile glass panels in balconies. 11 Otto Brandtin tie, helsinki; Photo by J. T. Cooke; June 2018 © J.T. Cooke. *Note the mobile glass screens above the solid balustrade that create a protected space. The glazing is primarily provided for thermal comfort in winter.*

The smaller the opening, the less noise ingress, but also, lower rates of natural ventilation. If the mobile acoustic-barrier is entirely open to maximise natural ventilation noise ingress will increase. Because the user cannot pre-empt the noise of a passing truck, or loud noise, it is unlikely the screen will be in the required position before the noise event. The need for user intervention is likely to be particularly problematic at night if the acoustic-barrier screen is open for night-time cooling, but the occupants are trying to sleep.

FIGURE 81 – ACOUSTIC ABSORPTIVE PANEL TO THE SOFFIT OF A BALCONY TO REDUCE THE NOISE FROM AN ADJACENT TROLLEY-BUS STOP AT 66 SCHAFFHAUSERSTRASSE, WINTERHUR, SWITZERLAND; REPRINTED FROM WWW.BRUAG.COM, BY BRUAG, 2018 © BRUAG. *NOTE: THIS PROJECT FOR A SENIORS HOME BY SCHWYTER BENZ ARCHITEKTEN AG IN COOPERATION WITH HOLLENSTEIN ARCHITEKTEN. THE CONTEXT OF THE SOLUTION CAN BE INSPECTED IN THE ADDRESS ABOVE. THIS TYPE OF SOLUTION IS USEFUL FOR THE SITUATION NOTED IN FIGURE 39*

FIGURE 82 – DETAIL OF ACOUSTIC ABSORPTIVE PANEL TO THE SOFFIT OF A BALCONY TO REDUCE THE NOISE FROM AN ADJACENT TROLLEY-BUS STOP AT 66 SCHAFFHAUSERSTRASSE, WINTERHUR, SWITZERLAND; REPRINTED FROM WWW.BRUAG.COM, BY BRUAG, 2018 © BRUAG. *NOTE: THIS PROJECT FOR A SENIORS HOME BY SCHWYTER BENZ ARCHITEKTEN AG IN COOPERATION WITH HOLLENSTEIN ARCHITEKTEN. THE CONTEXT OF THE SOLUTION CAN BE INSPECTED IN THE ADDRESS ABOVE.*

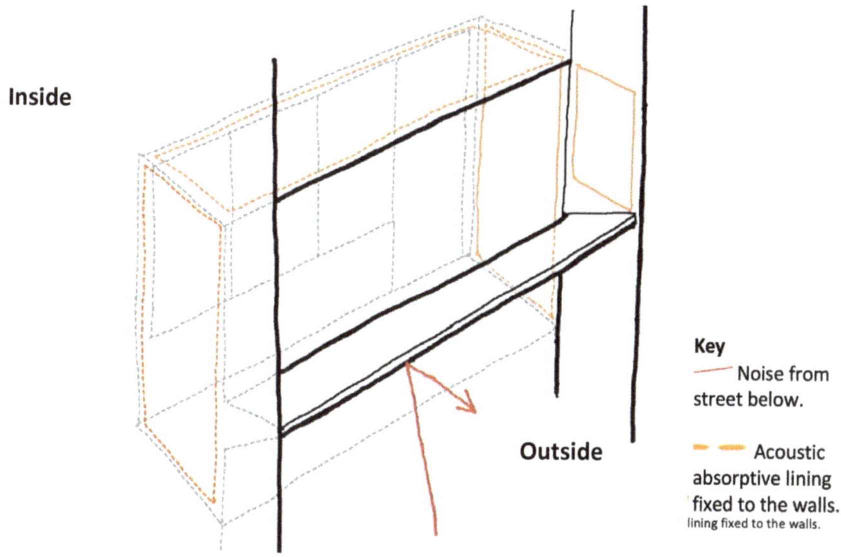

FIGURE 83 – AUGMENTED BALCONY BALUSTRADE TO PROVIDE NOISE REDUCTION; BY J.T. COOKE; BASED ON IMAGES COURTESY OF THE ENVIRONMENTAL PROTECTION DEPARTMENT AND THE HONG KONG HOUSING AUTHORITY (HKHA), THE GOVERNMENT OF HKSAR; FROM HTTPS://WWW.EPD.GOV.HK/EPD/INNOVATIVE/GREENY/ENG/BALCONY_ACOUSTIC.HTML; © 2016 ENVIRONMENTAL PROTECTION DEPARTMENT, AND HTTPS://WWW.HOUSINGAUTHORITY.GOV.HK/EN/ABOUT-US/PUBLICATIONS-AND-STATISTICS/HOUSING-DIMENSIONS/ARTICLE/20121109/INDEX.HTML#1; © 2011 HKHA; WEBSITE; MARCH 2021. *NOTE: LOCATED NEAR THE WEST KOWLOON MOTORWAY CORRIDOR, THE ANGLED GLASS PROFILE OF THE TOP OF THE SOLID BALUSTRADE IS DESIGNED TO REFLECT THE NOISE FROM THE MOTORWAY BELOW. THE ACOUSTIC ABSORPTIVE LINING IN THE BALCONY IS DESIGNED TO ABSORB AS MUCH SOUND ENERGY AS POSSIBLE OUT OF THE INCOMING AIR.*

As previously noted, the inclusion of an acoustically attenuated passive vent may help mitigate noise and provide for darkness in the case of night-time ventilative cooling. However, forming a barrier to the noise and careful site planning and internal space planning to locate balconies and bedrooms on the quiet facade would be a better solution, see Figure 94, Figure 101, and Figure 103. Designing the building to form a barrier to the noise source and receive natural ventilation from the quiet side of a building removes the need for occupant intervention to pre-empt the noise of a passing truck or loud noise. I will discuss designing to form a barrier to noise in the next section.

CHAPTER 6

FORMING BARRIERS TO NOISE

Tactics Primarily for Urban-Centre Edge, Consolidation and Ribbon Development areas

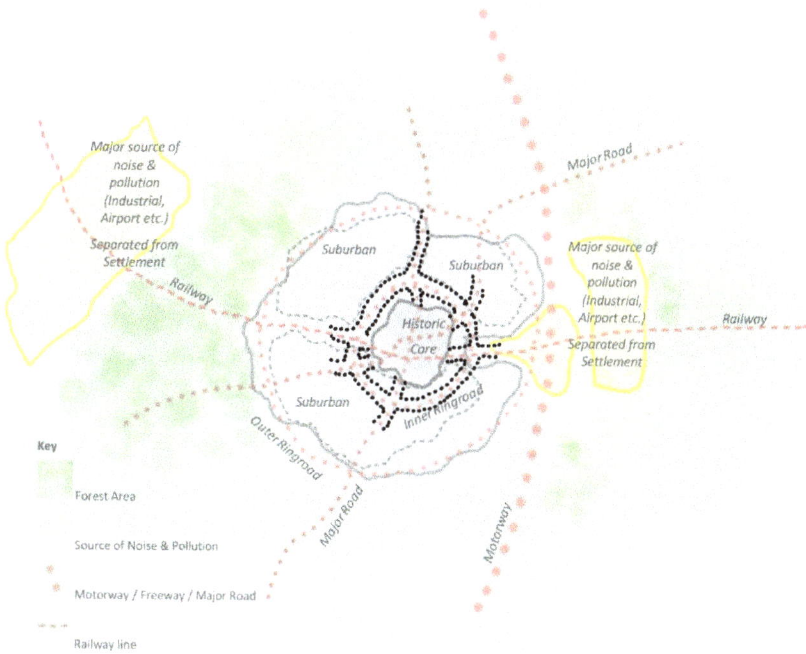

FIGURE 84 - A CARICATURE OF A SETTLEMENT FOCUSING ON THE AREA OUTSIDE THE URBAN-CENTRE, OR HISTORIC CORE WHERE RING ROADS AND ARTERIAL ROADS ARE OFTEN LOCATED, BY J. T. COOKE; JUNE 2019 © J.T.COOKE. *NOTE: IN SOME SETTLEMENT THERE CAN BE SUCCESSIVE RING ROADS.*

In some locations, and at severe levels of noise, absorption alone is less effective, and forming a barrier to the noise with the building becomes necessary. With the caricature of a settlement in mind illustrated in Figure 11 and Figure 84, moving from the denser urban-centre to the periphery of the urban-centre, there can be older suburban settlement forms. This part of a settlement may also have ring roads as sources of noise. Consolidation of the ribbon development and conurbation areas at the edge of the settlement centre is common along arterial roads and rail lines leading into a settlement. Some development can be noise-sensitive land uses such as housing. The proximity of noise-sensitive land-uses, including dwellings to the arterial motorways and railway lines, creates noise problems for residents. Planners, architects, designers, engineers, and developers need to work harmoniously and deploy solutions for this type of context in a settlement.

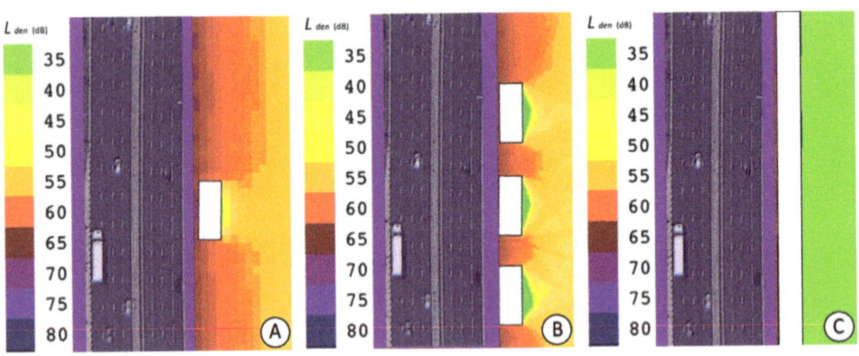

FIGURE 85 – DIAGRAM ILLUSTRATING THE IMPORTANCE OF CREATING A CONTINUOUS 'STREETWALL' BARRIER TO NOISE; BY J. T. COOKE; DEC 2020; BASED ON ACOUSTIC MODELLING OUTPUT FROM NOISEMAP DB; WWW.MASENV.CO.UK/DBMAP. *NOTE: THE IMPORTANCE OF DIFFERENT BUILDING LAYOUTS AND LEVELS OF CONTINUITY OF BARRIER FOR NOISE PROPAGATION. A – DETACHED BUILDING, 30M LONG, 15M WIDE & 15M HIGH, B – SEVERAL BUILDING WITH 15M SEPARATION BETWEEN THEM, C-CONTINUOUS 'STREETWALL' BUILDING. THE CONTINUOUS 'STREETWALL' HAS SIGNIFICANTLY LOWER NOISE LEVELS ON THE QUIET SIDE.*

Using absorption as a tactic becomes less effective with increased noise levels from sources like ring-roads and creating a barrier to the noise is a more effective solution. When creating a barrier to noise Figure 85 illustrates the importance of creating a continuous barrier to a noise source, and how significant the sound propagation can be depending on the size of the gap in the barrier. In the following section, the reader can inspect several examples of barrier building and structure solutions in the urban-centre edge ,and conurbation areas of settlements. Barriers can be designed in many ways and categorised into two main types: Noise barrier structures, and Noise barrier buildings

6.1 NOISE BARRIER STRUCTURES

Unoccupied structures

A noise-barrier-structure is a stand-alone structure that is specifically designed to deflect and block noise and are uninhabited. A barrier structure can be created in many ways depending on the specific constraints of the site. A barrier structure will usually be a solution where a noisy road passes through a noise-sensitive area, or as an infill structure inserted near an existing building to reduce noise for nearby residents. The design of the barrier can range from simple earth berms with fencing, to more complex systems, as illustrated in the examples in this section.

For typical road noise barriers, the U.S. Department of Transportation, Federal Highways Administration notes that '[I]in most cases, the maximum noise reduction that can be achieved by a barrier is 20 dB(A) for thin walls and 23 dB(A) for berms' [8] (2017, p. 1), and this is reflected in ISO9613.2 (see Figure 86), however; more complex structures can achieve more significant reductions.

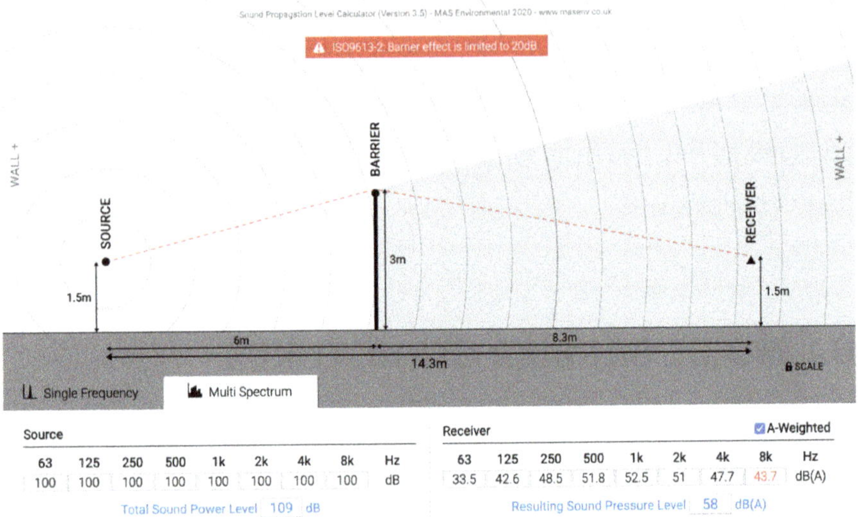

FIGURE 86 - DIAGRAM OF A SIMPLE NOISE BARRIER WITH CALCULATION; BY J.T.COOKE; DEC 2020; BASED ON ACOUSTIC MODELLING OUTPUT FROM MAS ENVIRONMENTAL NOISETOOLS; HTTPS://NOISETOOLS.NET/NOISECALCULATOR2

To reduce diffracted sound passing over the barrier building, or a perimeter block, the design of the top or side of any wall can be further augmented to improve the performance of the barrier (see Figure 35 D).

> Studies have shown that a T-profile top barrier provides insertion losses comparable to a conventional top barrier when the difference in their heights is equal to the width of the T-profile top. When the two barriers are the same height, the T-profile top barrier has been shown to provide an additional 2.5 dB(A) insertion loss over the conventional top barrier [9]

(U.S Department of Transportation; Federal Highway Administration, 2017, p. 2).

The inclusion of absorptive material in the design of the barrier apex has also been shown to assist performance. Designers could also apply designers logic to live-in sound barrier buildings. Designers could detail the parapet and corner of the buildings to improve the performance of the barrier building further.

Readers can investigate some examples of noise barrier structures below:

6.1.1 LEINSTER GARDENS, FACADES EXAMPLE

The false house facades of 23 and 24 Leinster Gardens, Bayswater, London were not built to be noise barriers, although they act as a noise barrier. The false fronts replace two houses demolished for the construction of the Metropolitan railway line in the 1860s. Originally built for steam engines which created a lot of noise and smoke, the facades formed a barrier to noise and smoke and continued the elegant streetscape of Leinster Gardens.

Readers can inspect the context of the project at the following coordinates:

51°30'45.4"N 0°11'01.6"W

and the following address:

23 and 24 Leinster Gardens, Bayswater, London W2 3AN, UK

and the following links:

Bing maps showing the rear of the facade and railway line open to the air:

www.bing.com

6.1.2 THEODOR-KÖRNER HOUSE EXAMPLE

The noise barrier structure in Vienna designed by Treusch Architecture and completed in 2007 is an example of a noise-barrier-structure inserted into an existing context. The site plan of an existing apartment complex included courtyard areas open on one side to a noisy motorway. The designers inserted the noise barrier structure of photovoltaic coated glass in the open section of the courtyard to reduce the traffic generated noise. Figure 85 illustrated the importance of creating a continuous barrier to block out noise, and the Theodor-Körner noise-barrier-structure does the same to block traffic noise and protect the public and private areas of the apartment complex. Readers can inspect the context of the project at the following coordinates:

48°10'49.3"N 16°21'25.9"E

and the following address:

Margaretengürtel, 1050 Vienna, Austria.

and the following links:

openstreetmap.org www.bing.com

6.1.3 A2 MOTORWAY UTRECHT EXAMPLE

The building is beside the A2 motorway (Ring) in Utrecht, in the Netherlands. The majority of the structure is a noise barrier structure, and so it is included as a structure rather than a building. One small section of the noise barrier consists of a car showroom integrated into the design. The inclusion of an integrated revenue-generating building offsets the cost of building the noise barrier. Designed in 2006 by ONL innovation studio and led by Kas Oosterhuis and Ilona Lénárd the innovative system creates an exciting solution to the problem of noise from the adjacent motorway and includes a car showroom. Readers can inspect the context of the project at the following coordinates:

52°06'48.5"N 5°02'36.2"E,

and the following address:

Proostwetering 53, 3543 AC Utrecht, Netherlands

and the following links:

openstreetmap.org

www.bing.com

6.2 NOISE BARRIER BUILDINGS

Occupied structures

A noise barrier building is an inhabited structure that blocks noise and is more often a commercial building rather than a residential building. A noise-barrier-building is commonly a less noise-sensitive use, such as a commercial building. In some cases, the noise-barrier-building can be a residential building, but designers must carefully design them to provide suitable habitat for the occupants. The space planning of noise barrier buildings is critically important, and the reader can see some examples of good space planning in the section on *Internal Space Planning*.

Readers can investigate some examples of noise barrier buildings below:

6.2.1 THE WALL SHOPPING CENTRE EXAMPLE

The Wall shopping centre is also adjacent to the A2 motorway (Ring) in Utrecht, in the Netherlands. This noise barrier building is 800m long and located nearby the noise barrier structure mentioned previously. The building is both a shopping centre and a barrier to noise from the adjacent motorway. Designed in 2010 by studio VVKH, the building blocks the traffic noise from reaching an extensive housing development beyond the building. This project is an example of inserting a less noise-sensitive land-use (retail) between the noise source (the motorway) and the noise-sensitive land-use (the housing). In addition to the land-use planning solution of creating a buffer, the building design provides a continuous noise barrier. The new housing built behind the shopping centre will need access to amenities and retail, and as long as it is a walkable distance, the solution has multiple benefits.

Readers can inspect the context of the project at the following coordinates:

52°06'25.1"N 5°03'25.7"E

and the following address:

Hertogswetering 183, 3543 AS Utrecht, Netherlands

and the following links:

Google.com/maps   openstreetmap.org

6.2.2 UNIVERSITY OF SYDNEY EXAMPLE

The Warren Centre building houses the Faculty of Engineering & School of Information Technologies at the University of Sydney. Completed in 2006 by Architects Francis-Jones Morehen Thorp, the construction creates a protective glass shield for the Centre and the areas behind the building to the south. The design responds to a challenging sloping site exposed to noise and air pollution from a busy road and junction. As illustrated in Figure 87 , it incorporates a noise-barrier glass curtain wall façade to reflect noise from the busy road. The glazed curtain wall façade protects the rooms in the Warren Centre from noise and extends beyond the edge of the building to protect the forecourt of the Seymour Centre from the traffic generated noise from Cleveland Street.

FIGURE 87 - THE WARREN CENTRE, FACULTY OF ENGINEERING AND SCHOOL OF INFORMATION TECHNOLOGY, UNIVERSITY OF SYDNEY, DESIGNED BY FJMT; PHOTO BY J. T. COOKE; JUNE 2019 © J.T.COOKE. *NOTE: THE GLAZED SCREEN ACTS AS A BARRIER TO THE NOISE FROM THE BUSY ROAD.*

The clever use of a glazed curtain wall allows a visual connection to the street while protecting the occupants from the traffic generated noise. The glazed curtain wall cantilevers off the building, leaving an acoustically attenuated interstitial space that includes landscaping to improve the outlook from the offices. Passing through the barrier is an audible journey from the soundscape of the busy road to the calmness of the protected courtyard. The glazed curtain wall facade solution is similar to that used by Treusch Architecture for the Theodor-Körner House example in Vienna; however, the Theodor-Körner House solution is purely an uninhabited facade inserted into the gaps in an existing apartment complex and not integrated into the new building design as in this case.

Readers can inspect the context of the project at the following coordinates:

33°53'16.7"S 151°11'38.0"E

and the following address:

School of Information Technologies, J12, The University of Sydney NSW 2006, Australia.

and the following links:

openstreetmap.org

6.2.3 BOURKE STREET APARTMENTS EXAMPLE

Figure 88 – 2016 NSW AIA Multiple Housing Architecture Award Barrier Building, Designed by MacGregor Westlake; Photo By J. T. Cooke, June 2019 © J.T.Cooke. *Note: The housing is located in a difficult site adjacent to a motorway and forms a barrier to the noise from the traffic.*

This noise barrier building is a residential apartment building and uses a colourful wall to block noise from the motorway entering the homes. The adjacent M1 motorway carries more than 40,000 vehicles Annual Average Daily Traffic (AADT) and up to approximately 90,000 Average Daily Traffic Count and passes through part of the historic core of Sydney. The building also uses space planning to reduce noise exposure locating less noise sensitive rooms on the motorway side of the building. Completed in 2016 by McGregor Westlake Architecture, the award-winning building responds to a challenging site and a requirement that all apartment buildings be naturally ventilated and naturally cross-ventilated. The apartment buildings must maintain an internal LAeq decibel level of not more than 35dB(A) in any bedroom between 10 pm to 7 am, and not more than 40dB(A) at all times in all other rooms except kitchens, bathrooms or hallways, with windows open for ventilation.

Readers can inspect the context of the project at the following coordinates:

33°52'15.2"S 151°13'05.7"E

and the following address:

13-15 Griffiths St

Woolloomooloo NSW 2011, Australia.

and the following links:

openstreetmap.org

Readers should remember the example of Munich from section 3 of the book and the ring roads around the city. Some excellent examples of noise barrier buildings have been completed in Munich around the Innsbrucker Ring Road to improve living conditions for residents. The City of Munich used progressive policy and design competition to spur the development of many excellent examples and illustrated them in a publication called 'Peace at last; new dwellings on the middle ring road'. A funding program called 'dwelling on the ring' provided grants for noise protection measures in residential buildings to improve the quality of life for residents. The Innsbrucker Ring Road carries approximately 64,000 vehicles per day and noise levels were measured at 71 dB(A) during the day and 61 dB(A) at night. The design innovations and addition of the in-fill noise barrier building significantly improved the noise levels for both the existing and new housing blocks. (City of Munich: Department of Urban Planning and Building Regulations, 2012).

Readers can investigate the following examples:

6.2.4 AGE CARE HOUSING, MÜNCHEN EXAMPLE

FIGURE 89 - PHOTO OF THE NEW NOISE BARRIER HOUSING BLOCKS FROM THE NOISY SIDE ON THE INNSBRUCKER RING 70 & 72 IN MUNICH. COURTESY OF KRIEGER ARCHITEKTEN BDA, BY KRIEGER ARCHITEKTEN BDA; 2008 © KRIEGER ARCHITEKTEN BDA, SAMERBERG: WWW.KRIEGER-ARCHITEKTEN.DE. *NOTE: THE ELEVATION FACING THE MOTORWAY USES PAINTED ACCESS CORRIDORS TO CREATE A DYNAMIC PRESENCE AND AS A BUFFER TO THE TRAFFIC NOISE.*

FIGURE 90 - PHOTO OF THE ELEVATION OF THE NOISY-SIDE OF THE NEW NOISE BARRIER HOUSING BLOCKS FROM THE NOISY SIDE ON THE INNSBRUCKER RING 70 & 72 IN MUNICH. COURTESY OF KRIEGER ARCHITEKTEN BDA, BY KRIEGER ARCHITEKTEN BDA, 2008 © KRIEGER ARCHITEKTEN BDA, SAMERBERG: WWW.KRIEGER-ARCHITEKTEN.DE. *NOTE: THE COLOURFUL ELEVATION AND WINDOW OPENINGS IN THE GLASS FAÇADE FOR VENTILATION.*

Age Care housing was used as an infill development to create a barrier to noise from the adjacent Innsbrucker Ring 70 in München, Germany. As illustrated in Figure 89 to Figure 91 the new infill addition to the existing housing blocks create a barrier to the noise from the motorway and a dynamic elevation. The original housing blocks were perpendicular to the road, which allowed the space between the buildings to fill with traffic generated noise, as illustrated in Figure 93. The insertion of the new aged-care housing blocks significantly reduced the noise levels from the road for the existing and new residents. Completed in 2010 by Krieger Architekten BDA the project received the Honorary Prize for Good Housing, Aged Care Housing, and exemplary renovation from the City of Munich.

FIGURE 91 - GENERAL ARRANGEMENT PLAN OF ADDITION OF NOISE BARRIER HOUSING BLOCKS AND REFURBISHMENT OF EXISTING HOUSING ON THE INNSBRUCKER RING 70 & 72 IN MUNICH. COURTESY OF KRIEGER ARCHITEKTEN BDA, BY KRIEGER ARCHITEKTEN BDA, 2008 © KRIEGER ARCHITEKTEN BDA, SAMERBERG: WWW.KRIEGER-ARCHITEKTEN.DE. *NOTE: THE NEW AGED-CARE HOUSING, CIRCLED IN RED, FORMS A BARRIER TO NOISE FROM THE MOTORWAY AND CREATES QUIET COURTYARDS THAT FORMERLY WERE SIGNIFICANTLY AFFECTED BY TRAFFIC NOISE. THE NEW BARRIER BUILDING HAS ALSO CONSIDERABLY REDUCED TRAFFIC NOISE, REACHING THE EXISTING HOUSING IN GREY.*

When creating a barrier building, particularly a residential barrier building, it is crucially important to use the correct space planning to prevent the occupants of the barrier building from being subjected to noise. The author will discuss this further in the next section. The space-planning of the new aged-care infill blocks is an excellent example of using the location of corridors, kitchens, and bathrooms in the plan as a barrier to noise from the busy road. As illustrated in Figure 92, the glazed access corridor has been used as a buffer to traffic noise and a colourful active facade, as illustrated in Figure 89 and Figure 90. Within the apartments, the kitchens and bathrooms provide further buffering of any noise entering the dwelling, as illustrated in Figure 92. Bedrooms and balconies are positioned as far from the noise source as possible, and all access to light and ventilation is from the quiet courtyard side of the building.

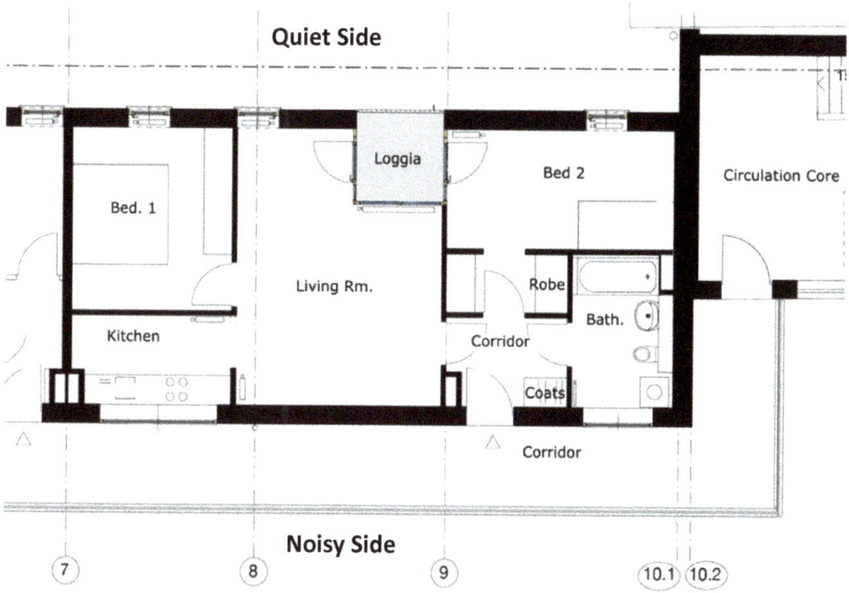

Figure 92 - A detailed plan of a two-bedroom apartment in the noise barrier housing blocks on the Innsbrucker Ring 70 & 72 in Munich. Adapted from plan courtesy of Krieger Architekten BDA, by Krieger Architekten BDA, 2008 © Krieger Architekten BDA, Samerberg: www.krieger-architekten.de. *Note: The glazed access corridor is used as a buffer to traffic noise, and the kitchen and bathroom further buffer noise where small windows are located in the external wall. The bedrooms are located at the furthest point from the noise source, and all natural-ventilation openings to the bedrooms and living area are from the quiet courtyard side. The Loggia on the quiet side could also act as a sound absorption chamber, reducing any social noise from the courtyard.*

Figure 93 – Diagram of approximate analysis of before and after insertion of the barrier building; by J. T. Cooke; based on acoustic modelling output from Noisemap Db; www.masenv.co.uk/dbmap; Dec 2020. *Note: There is a significant reduction in the traffic noise level within the newly formed courtyards.*

Readers can inspect the context of the project at the following coordinates:

48°07'18.7"N 11°37'13.7"E

and the following address:

Innsbrucker Ring 70

81671 München

Germany

and the following links:

openstreetmap.org

www.bing.com

6.2.5 AFFORDABLE HOUSING MÜNCHEN, EXAMPLE

This award-winning alterations and addition of affordable housing to an existing apartment building complex was completed in 2012 by Felix + Jonas Architekten BDA. Similar to the previous aged-care housing example, this project also inserts infill affordable housing blocks as a barrier to the noise from the ring road. As with the aged-care example, the original housing blocks were perpendicular to the road, which allowed the space between the buildings to fill with traffic-generated noise. The new infill blocks created two large quiet courtyards and a third smaller courtyard, and formed a barrier to the traffic-generated noise. Rather than a blank façade facing the ring road, the design creates a vibrant and colourful façade and includes some windows for light and passive street surveillance.

Readers can inspect the context of the project at the following coordinates:

48°07'00.7"N 11°37'05.4"E

and the following address:

Zoenedinger Strasse 12-38, München
Germany

and the following links:

openstreetmap.org   www.bing.com

Creating a barrier with the building layout and site planning can be done using two primary urban forms at different scales, which are linear and courtyard shapes, as illustrated in Figure 94. As discussed in this section, the continuity of a barrier form is critical in its effectiveness as a noise barrier. In the following section, I will discuss the two primary urban forms for designing near a noise source, and the reader can examine some examples of each type and their suitability for the context.

6.3 SITE PLANNING

The layout of buildings on a site

As illustrated in Figure 93, a noise barrier performs better if it is a continuous barrier. In urban design and planning, two primary urban forms can create a constant noise barrier: linear forms and courtyard forms, as illustrated in Figure 94. The objective of the urban layout is to create a defensive streetwall to the noise and a protected, quiet courtyard or garden side to the building. At an early stage in the design process, a well-considered design can reduce the potential exposure of inhabitants to noise and provide a more economical solution than using mechanical or acoustic devices later in the design process to improve a fundamentally flawed design. The arrangement of buildings on a site can be an effective way of creating a noise barrier, but designers should be careful when making a barrier building not to reflect noise onto neighbouring properties. Sound absorption must be used in combination with layout and barrier solutions to mitigate the reflected noise. In designing barrier buildings, designers must not propose a poor urban design outcome, such as blank facades, without passive street surveillance. As the previous examples illustrate, it may be possible to include double or triple glazed windows on the noisy side to provide street surveillance and light, but not natural ventilation and noise ingress. Many of the internal space planning examples in the next section use sealed double glazing or triple glazed windows on the noisy side to provide street surveillance and more active, interesting facades. Still, the solution will depend on the specific context.

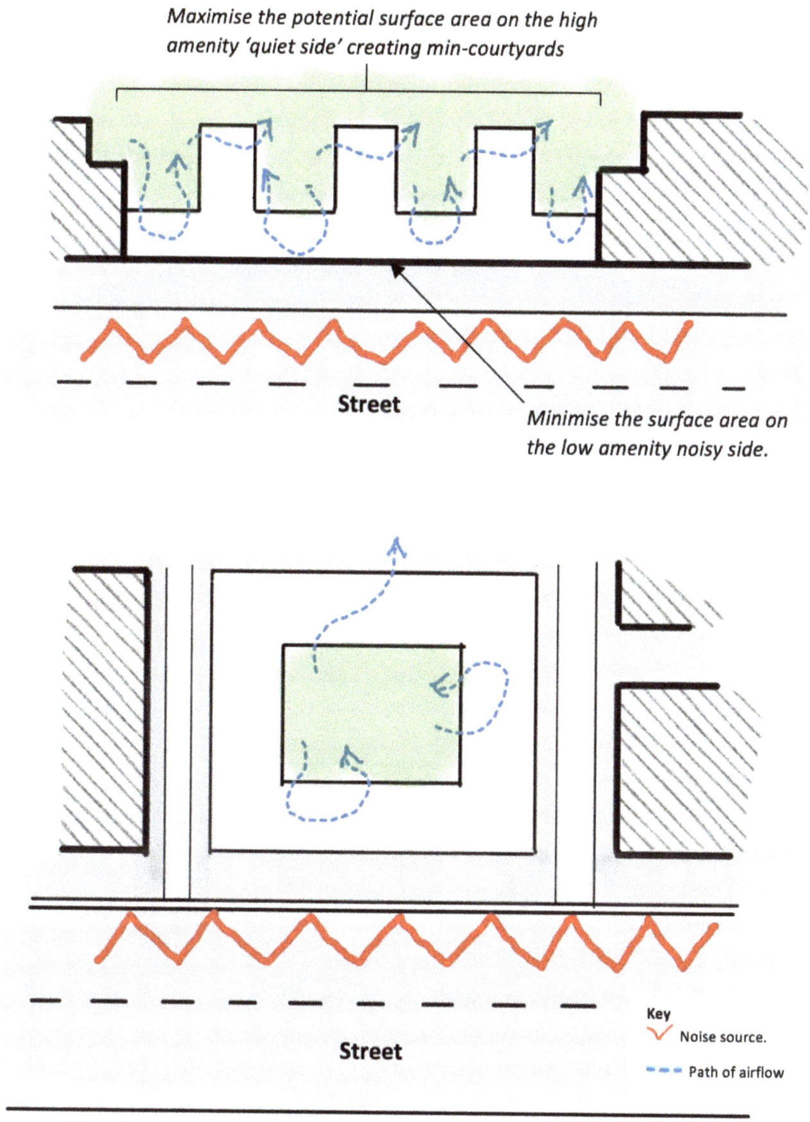

FIGURE 94 – PLAN OF LINEAR BARRIER BUILDING AND COURTYARD BUILDING; BY J. T. COOKE; JUNE 2018, © J.T.COOKE. *NOTE: THE TWO MAIN TYPES OF CONTINUOUS BARRIER URBAN FORM: A LINEAR FORMS AND A COURTYARD FORM. THE FLOORPLATE FACING THE NOISE SOURCE WILL BE THINNER BECAUSE IT WILL BE SINGLE-SIDED VENTILATION FROM THE QUIET SIDE. SINGLE-SIDED VENTILATION IS LIMITED TO ABOUT 6-8 M.*

As discussed in *Chapter 1 Relevant Historical Precedents*, there are many examples of urban form in higher-density settlements that consist of perimeter block buildings arranged around courtyards.

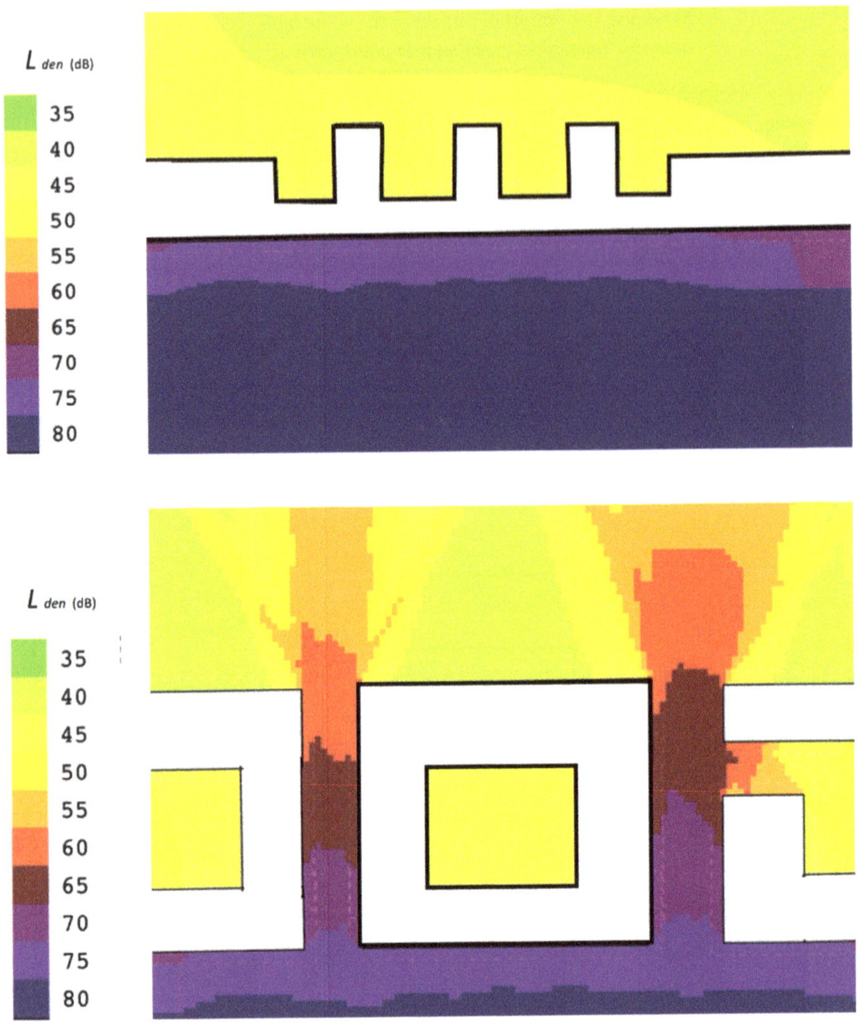

Figure 95 – Diagrammatic analysis of linear barrier building and courtyard building, by J. T. Cooke; based on acoustic modelling output from Noisemap Db; www.masenv.co.uk/dbmap; Dec 2020. *Note: The barrier building and Courtyard form both significantly reduce the sound level on the quiet side. Further acoustic attenuation on the quiet side may be needed to maintain the quiet quality of the quiet side, and acoustically attenuated vents and windows may be required to achieve required sound levels with windows and vents open for natural ventilation. A barrier wall with solid access gates similar to traditional courtyard access-lanes could also be included at the entry to the side-streets to prevent sound travelling down.*

The perimeter block forms a barrier to the street noise and shelters the courtyard. As the streets in urban-centres became ever more dominated by machine noise, inner courtyards took on a peaceful, restorative function, and providing a quiet side for dwellings has public health benefits, as noted in the section about *Noise and Public Health*. Access to a 'quiet-side' suggests that dwellings should be dual-aspect, with one aspect facing a quiet side, such as a courtyard form (see Figure 94 and Figure 100). Forming perimeter blocks around a quiet courtyard can be a good initial step to provide a quiet side. As noted in the section about *Noise and Public Health* it is essential that the quiet-side provides great amenity and is truly quiet. Because noise can reverberate around a courtyard, as shown in Figure 96, designers must design the courtyard to be a quiet space. To achieve a quiet courtyard, a designer can use sound-absorbing material and acoustic vents & windows to mitigate against social noise created and minimise sleep disturbance. Using perimeter block forms in conjunction with considered space planning and noise attenuation can offer a restorative place where inhabitants can escape noise (see Figure 94 and Figure 100).

A perimeter-block urban-form (courtyard building) provides an excellent basis for good acoustic design and human habitat within urban centres. It may be helpful to watch the video below for further explanation:

YouTube Channel

FIGURE 96 - DIAGRAMMATIC ANALYSIS OF COURTYARD BUILDING WITH NOISE IN THE COURTYARD, BY J. T. COOKE; BASED ON ACOUSTIC MODELLING OUTPUT FROM NOISEMAP DB; WWW.MASENV.CO.UK/DBMAP; DEC 2020. *NOTE: IF THERE IS NOISE IN THE COURTYARD, THE SHAPE OF THE ENCLOSURE CAN TRAP THE NOISE REVERBERATING IT AROUND AND SIGNIFICANTLY REDUCING THE BENEFIT OF A QUITE-SIDE. DESIGNERS SHOULD DESIGN THE COURTYARD TO ABSORB AND DEPLETE SOUND ENERGY, AND DESIGNERS COULD USE DOUBLE-FRAME WIDOWS OR ACOUSTIC VENTS TO MITIGATE ANY SLEEP DISTURBANCE FROM SOCIAL NOISE.*

In the author's experience, designing a shallower floorplate depth for any building facing a noise source will allow for the noisy facade to be acoustically sealed while providing single-sided natural ventilative cooling from the quiet courtyard side, see Figure 100. The noisy facade could still contain double or triple glazed windows to provide light but should not be used for ventilation. Designers should limit the overall envelope depth of buildings parallel to a noise source to approximately 10 metres, and living rooms, kitchens and bedrooms should not be further than 6 to 8 metres from the natural ventilation openings on the quiet side, but it will depend on the specific requirements of the project. Readers can investigate the following good examples of site planning to create barrier buildings from Germany:

6.3.1 GEORG WILHELM HOUSE EXAMPLE

Designed in 2011 by Czerner Göttsch Architekten, the international award-winning masterplan design used linear and courtyard forms to respond to a challenging site and create a barrier to the noise in an industrial area. The initial stage of the project on the corner of Kurdamm and Georg-Wilhelm Strasse is complete, but the site to the south has yet to be built. Using a courtyard form for the initial stage is a clever tactic when building in a noisy area because the courtyard form can provide a self-contained continuous noise barrier. The completed housing creates a protected, quiet courtyard form to shelter residents from surrounding noisy uses. When finished, the linear streetwall of dwellings to the south will form a shield to defend the area to the east from industrial noise, as shown in Figure 97.

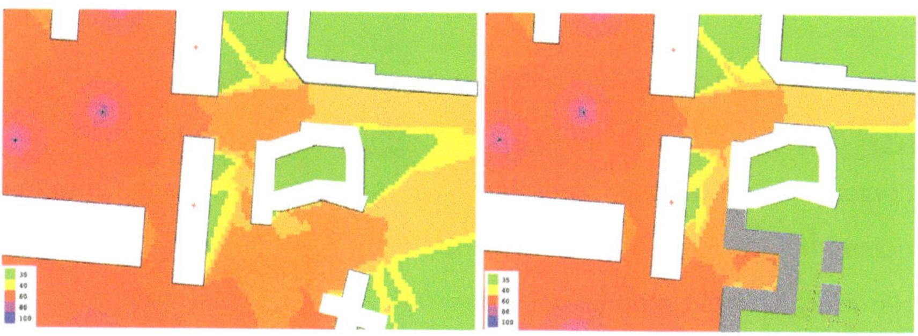

FIGURE 97 - DIAGRAMMATIC ANALYSIS OF A COURTYARD BUILDING AND STAGE 2 BARRIER BUILDING, BY J. T. COOKE; BASED ON ACOUSTIC MODELLING OUTPUT FROM NOISEMAP DB; WWW.MASENV.CO.UK/DBMAP; DEC 2020. *NOTE: THE NOISE LEVELS ARE SIGNIFICANTLY REDUCED WITHIN THE COURTYARD AND AFTER THE INCLUSION OF THE STAGE 2 BARRIER BUILDING.*

Readers can inspect the context of the project at the following coordinates:

53°29'45.5"N 9°59'28.6"E

and the following address:

Kurdamm 2, 21107

Hamburg,

Germany

and the following link:

openstreetmap.org

6.3.2 CITY GARDENS LOKSTEDT, EXAMPLE

The City Gardens, Lokstedt, site 1, project was completed in 2013 by LRW Architects & City Planners. The apartment building and childcare centre in Grandwegs Strasse, Hamburg, is an excellent response to the noise from the adjacent sports complex. The adjacent Grandweg Hamburg Sports Complex generates nuisance noise, and the apartment building forms a continuous protective shield for occupants and neighbouring dwellings, as illustrated in Figure 98. As with many barrier building plans, the development maximises the external building envelope of the courtyard side by creating projections into the inner courtyard to maximise the surface area for apartments on the quiet side. As illustrated in Figure 99, the apartments fold around the circulation cores forming projections into the courtyard and increasing the surface area on the quiet side for bedrooms and balconies.

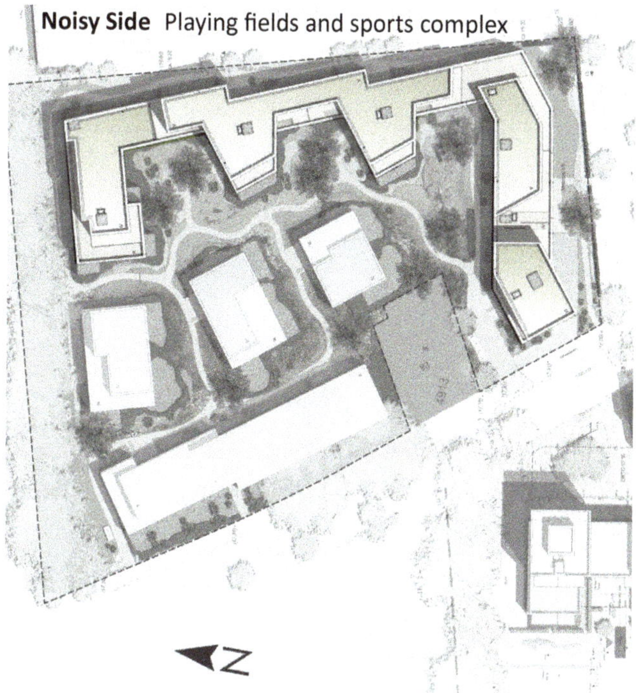

Figure 98 - New housing with daycare facilities in Lokstedt Hamburg. Courtesy of Loosen, Rüschoff + Winkler PartG mbB Architekten und Stadtplaner (LRW), by LRW, 2013, Hamburg: © LRW. www.lrw-architekten.de *Note: the continuity of the barrier building forming a courtyard and the increased surface area of the barrier building on the quiet side.*

Figure 99 - New housing with day-care facilities in Lokstedt Hamburg. Courtesy of Loosen, Rüschoff + Winkler PartG mbB Architekten und Stadtplaner (LRW), by LRW, 2013, Hamburg: © LRW. www.lrw-architekten.de *Note: the narrow building depth of approximately. 9-10 meters of the dual aspect apartments facing the noisy side.*

Readers can inspect the context of the project at the following coordinates:

53°35'27.0"N 9°57'51.7"E

and the following address:

Grandweg 95

22529 Hamburg

Germany

and the following links:

openstreetmap.org www.bing.com

Overall, perimeter-block, courtyard urban-forms are an excellent response to noise; however, in the author's experience there are several issues with implementing them in practice:

> **They require large sites** to enable them. Where phased development is proposed requiring consolidation of multiple sites, they need government-power over landowners to consolidate small sites into larger ones, or hope that adjacent sites develop at the same time. In the authors

experience, achieving cooperation among landowners to agree to masterplans that span multiple sites is very difficult.

Planning controls use the boundary as their starting point often resulting in buildings that are objects in a site offset from the boundary.

They require a master-planned design-led approach, and by the time a project gets to the planning assessment stage, money has been spent on design fees, and there is a reluctance to pay for re-design. At the planning assessment stage, it is too late to produce a master-planned urban form, and designs get pushed down the route of a quick-fix addition, and usually air-conditioning to fix a flawed design.

Designers' context analysis reflects the primacy of the visual sense and neglects the invisible acoustic context. It is not until a client pays for an acoustic analysis at the planning assessment stage that designers get the first assessment of the acoustic context, and unfortunately, a significant change to the design is unlikely because of the costs incurred to reach the planning assessment stage.

In the author's experience, it is possible, but difficult for planning controls to deliver perimeter-block urban-form, for infill development and it is more likely on larger sites, usually for re-development. It would be beneficial if land-law and planning controls made it more straightforward to achieve courtyard, perimeter block buildings, at least in the urban-centre of settlements, and near noise sources such as busy roads and railways. Even when planning authorities have set built-form envelopes, there is a need for correct space planning for noisy contexts to be carried through into the detailed design stage, but sadly this is often lacking in practice.

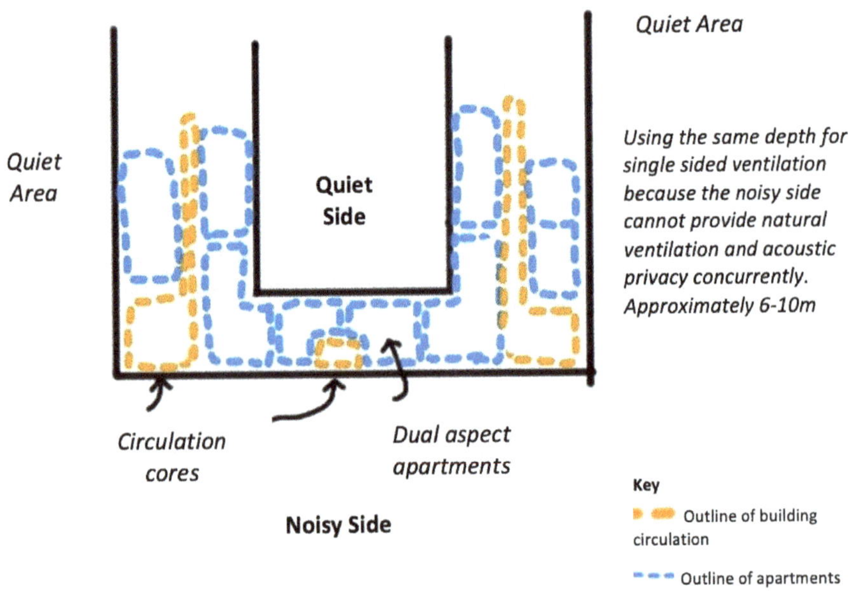

FIGURE 100 – PLAN OF COURTYARD BUILDING SHOWING APARTMENT LOCATIONS; BY J. T. COOKE; JUNE 2018, © J.T.COOKE. *NOTE: THE THINNER FLOORPLATE DEPTH FACING NOISE SOURCE (ROAD) TO PROVIDE FOR WHAT WOULD BE SINGLE-SIDED VENTILATION IF THE NOISY SIDE IS SEALED.*

It is equally vital that buildings create an excellent urban design and a sense of place and enclosure. The creation of a streetwall with either courtyard or linear barrier building urban forms can provide a good acoustic environment and a sense of enclosure in the street. A sense of enclosure in the public room that is the street can create a sense of place, and particularly in an urban centre in a settlement. Baron Hausmann's famous boulevards in Paris made a wonderful sense of enclosure of the street and potentially provided for a quiet side to the rear of the build. Figure 101 to Figure 105 illustrate space-planning for apartment design that can provide passive surveillance of the street and acoustic privacy & natural ventilation for occupants. To create suitable human habitat, good urban design, and places people want to live requires thoughtful and considered design and planning.

6.4. INTERNAL SPACE PLANNING

The arrangement of rooms within a building

At a more detailed design stage, the interior space planning of a building can be used to form a barrier to noise. The principle is to put less noise-sensitive rooms (uses) as a buffer to mitigate noise, which can be laundries, kitchens, dining rooms, corridors or bathrooms, as illustrated in Figure 101 to Figure 103. From the author's experience, most apartment plans are designed from the inside out and are stuck in a time when buildings were designed as objects and treated their site as a *tabula rasa*. The author believes that good design can solve problems and make people's living conditions better, but designers must begin from the position of creating good a human habitat and responding to the uniqueness of the context. In the author's experience, most apartment plans submitted for planning assessment do not respond to context, nor acoustic-context, and are mainly regurgitations of standard plans repeated to maximise *yield*. The space planning of buildings and the shape of rooms in noisy environments can have a significant impact on the quality of life of the inhabitants.

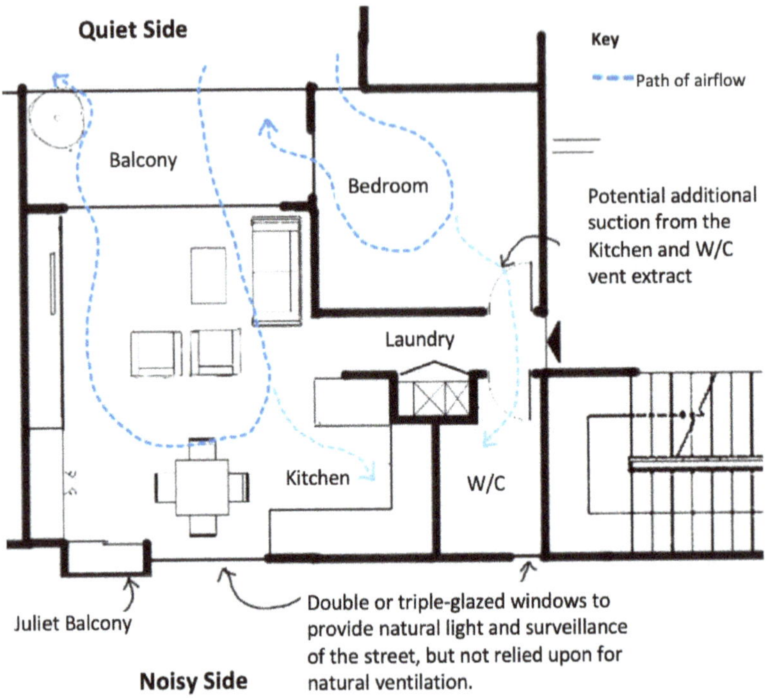

FIGURE 101 – DUAL-ASPECT APARTMENT PLAN THAT COMBINES PASSIVE SURVEILLANCE OF THE STREET, ACOUSTIC PRIVACY, AND NATURAL VENTILATION; BY J. T. COOKE; 2015 © J.T.COOKE. *NOTE: THIS PLAN ALLOWS FOR SINGLE-SIDED NATURAL VENTILATION FROM THE QUIET SIDE, AND THIS IS PARTICULARLY IMPORTANT IF USING A NATURAL VENTILATION NIGHT-COOLING STRATEGY TO COOL BEDROOMS AND LIVING AREAS. DEPENDING ON THE DECIBEL LEVEL OF THE NOISE SOURCE, THE OCCUPANT CAN USE SHORT-TERM, PURGE CROSS-VENTILATION BY OPENING THE DOUBLE-GLAZED BALCONY DOOR ON THE NOISY SIDE. HOWEVER, THIS SHOULD NOT BE RELIED UPON FOR COOLING, NOR VENTILATION WITHOUT ACOUSTIC VENTS, OR WINDOWS DESCRIBED PREVIOUSLY IN THIS BOOK. THE INCLUSION OF DOUBLE-FRAME ACOUSTIC WINDOWS, OR VENTS COULD ENABLE CROSS-VENTILATION THROUGH THE NOISY SIDE. THE BLUE ARROWS SIGNIFY AIRFLOW.*

With thought and a client who cares, innovative architects and designers can produce internal space planning designs that provide excellent design, good habitat, and a development for their client. If the correct building envelope has been set in the master planning stage, with a thinner floorplate on the noisy side, then the internal space planning at the detailed design stage is far more manageable. If too wide a floorplate has been assumed and set at the master planning or building envelope stage, it will be difficult for the designer to fill it later.

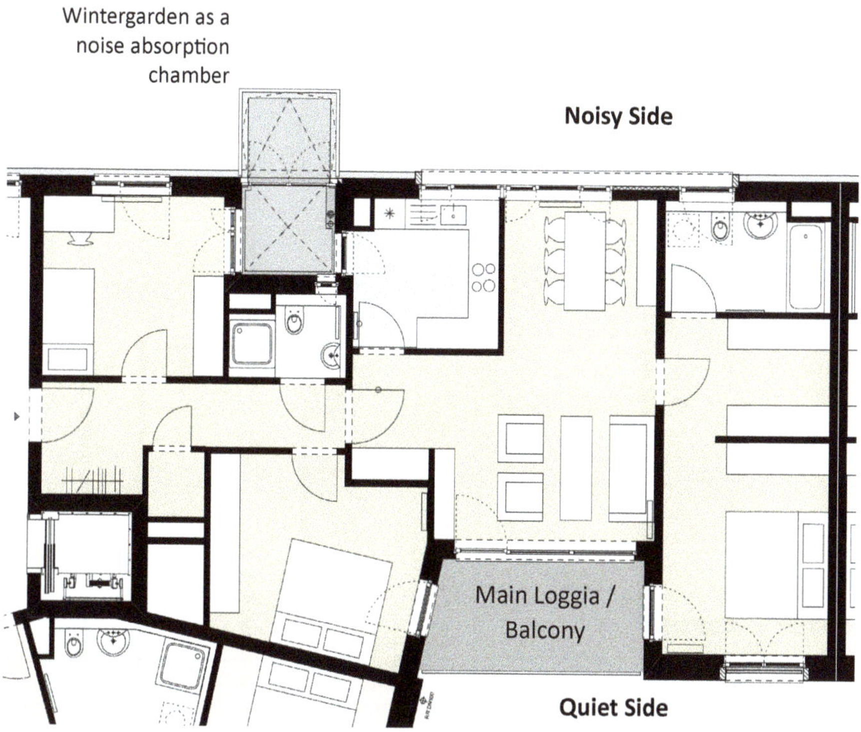

FIGURE 102 - NEW HOUSING WITH DAY-CARE FACILITIES IN LOKSTEDT, HAMBURG. COURTESY OF LOOSEN, RÜSCHOFF + WINKLERARCHITEKTEN UND STADTPLANER (LRW), BY LRW, 2013, HAMBURG, © LRW, WWW.LRW-ARCHITEKTEN.DE. *NOTE: THE KITCHEN AND BATHROOM FACE THE NOISY SIDE, AND A NOISE ABSORPTION LOGGIA CAN PROVIDE NATURAL VENTILATION WITH ACOUSTIC PRIVACY TO THE KITCHEN AND STUDY/GUEST BEDROOM. THE TWO MAIN BEDROOMS ON THE QUIET SIDE CAN BOTH RECEIVE NATURAL VENTILATION FROM THE MAIN LOGGIA / BALCONY. A NIGHT-COOLING STRATEGY CAN BE USED BECAUSE BEDROOMS ARE ON THE QUIET SIDE OF THE BUILDING WITH AIRFLOW PASSING THROUGH THE ACOUSTIC ATTENUATION OF THE MAIN BALCONY. NATURAL CROSS VENTILATION WITH ACOUSTIC PRIVACY IS POSSIBLE BECAUSE THE NOISE ABSORPTION LOGGIA ATTENUATES NOISE WHILE ALLOWING AIRFLOW TO PASS TO THE QUIET SIDE.*

When beginning space planning in a noisy context, consider and categorise room uses based on sensitivity to noise and locate noise-sensitive uses as far from the noise source as possible. It is as important to think about the ventilation openings required for each room, position them as far from the noise source as possible, and follow the principles discussed in the section locating ventilation openings and windows to mitigate noise as much as possible.

Figure 103 - A plan of a dual-aspect three-bedroom apartment that combines passive surveillance of the street, acoustic privacy, and natural ventilation, by J. T. Cooke; 2015 © J.T.Cooke. *Note: This plan is a derivation of the plan in Figure 101. The space-planning tactic in this plan stacks the bedrooms to create a castellated protrusion in the plan, as illustrated in Figure 94, which increases the surface area for natural ventilation. The occupant could use a Night-cooling strategy because the bedroom ventilation is from the quiet side. The designer could further reduce the possibility of noise-related sleep disturbance by specifying acoustic vents, or double window frame acoustic windows.*

The shape and location of the rooms can both be used to reduce noise levels when space-planning. Nijs et al. (2004) reported a difference of 10dB to the receiver between a rectangular room and a U-shaped room. The U-shaped room provides a quieter area in the corner farthest from the noise source, as illustrated in Figure 104. A designer should consider the source of noise and locate the ventilation openings in locations where the room shape produces a noise reduction. The room shape may be critical when designing studio apartments and other single-room dwellings.

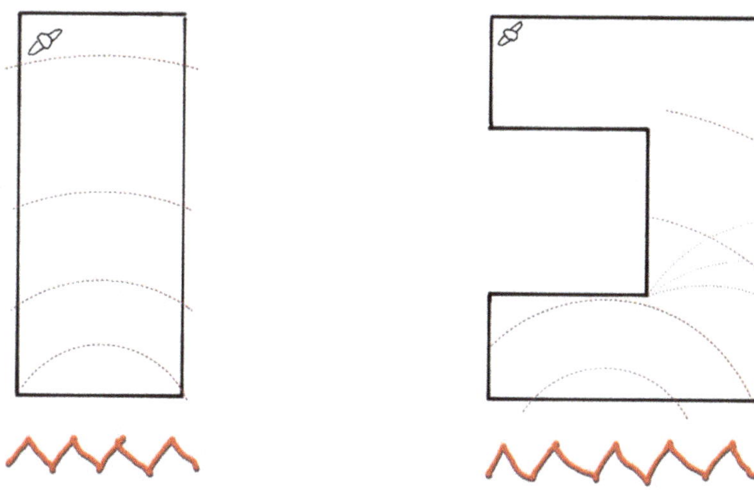

Figure 104 – Reduction in sound level because of room shape; by J. T. Cooke; 2018 © J.T.Cooke. *Note: Nijs et al. (2004) reported a difference of 10dB to the receiver between a rectangular room and a U-shaped room. The shape of the room can form a buffer to the noise. Consideration of room shape may be particularly important when designing studio apartments.*

The design and location of the bedroom is an obvious consideration in space planning in a noisy environment. Bluhm et al. (2004) found that, when bedrooms did not face the street, residents reported lower levels of sleep disturbance, and found that having access to a quiet facade reduced disruption to residents on average by 30-50%; and noted that apartment dwellers were more affected than house dwellers. (Bluhm, Nordling, & Berglind, 2004, pp. 43-49).

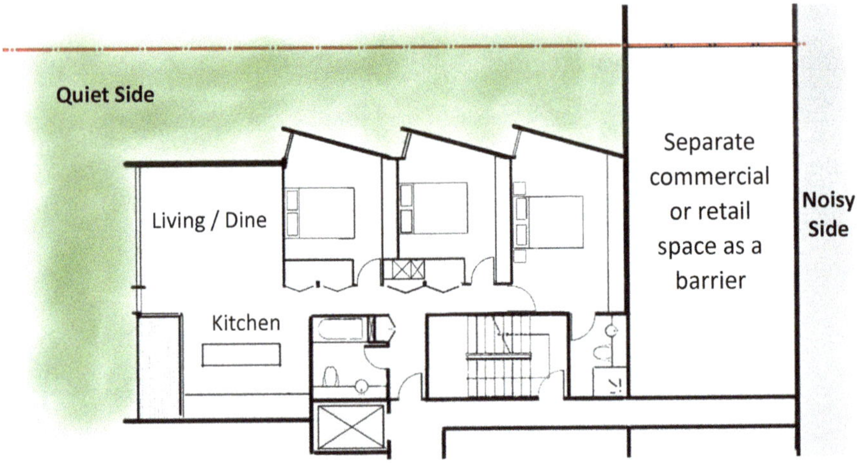

FIGURE 105 - A PLAN OF A THREE-BEDROOM APARTMENT THAT USES A COMMERCIAL OR RETAIL SPACE AS A BARRIER TO NOISE AND STACKS BEDROOMS, BY J. T. COOKE; 2015, © J.T.COOKE. *NOTE: THIS PLAN IS A DERIVATION OF FIGURE 103, BUT THE LIVING AREA IS MOVED TO THE END OF THE PLAN, AND A SEPARATE COMMERCIAL OR RETAIL SPACE IS USED AS A BARRIER TO NOISE. THE BEDROOMS ARE STACKED AND GAIN NATURAL VENTILATION FROM THE QUIET SIDE. THE LIVING AREA MAY BE ABLE TO ACHIEVE NATURAL CROSS VENTILATION. THE STACKING OF THE BEDROOMS ON THE QUIET SIDE CAUSES A PROTRUDING PLAN, AS ILLUSTRATED IN FIGURE 94.*

Basner and McGuire (2018, p. 38) noted, ‘There is biologic plausibility that chronic night-time exposure to relevant levels of noise can contribute to negative health consequences like a cardiovascular disease’ (MDPI licenced under CC BY). The appropriate design for a noisy context isn't a luxury; it can have health consequences for those who have to live in the buildings and is a public health issue. It may be that as we move into the future, some sources of noise in our habitat will reduce, but there will always be noisy contexts in settlements.

Greater attention should be paid to the design of dwellings in a noisy context, and strategies and tactics as illustrated in Figure 94, and

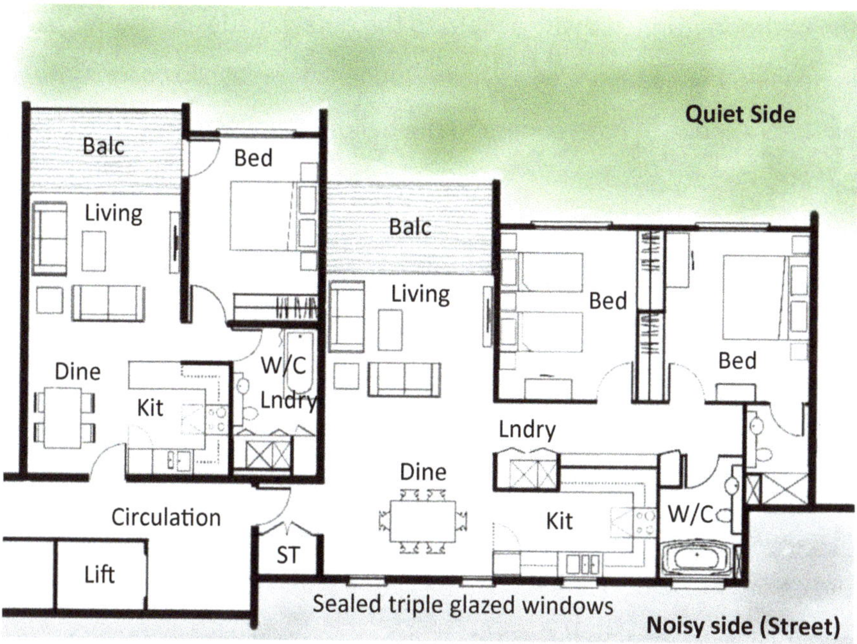

FIGURE 106 – DIAGRAMMATIC PLAN WITH FOUR APARTMENTS AROUND CORE ARRANGEMENT; BY J.T.COOKE; 2019, © J.T.COOKE. *NOTE: IN THIS PLAN ALL BEDROOMS RECEIVE NATURAL VENTILATION FROM THE QUIET SIDE AND LIVING AREA BALCONIES ARE ON THE QUIET SIDE. PASSIVE SURVEILLANCE TO THE STREET CAN BE PROVIDED BY THE SEALED TRIPLE GLAZED WINDOWS ON THE NOISY SIDE. THE PLAN CAN BE MIRRORED AND REPEATED TO CREATE A CONTINUOUS STREETWALL.*

Figure 100, to increase the surface area on the quiet side, stacking bedrooms on the quiet side as in Figure 103 and locating bedrooms on the quiet side as in Figure 101 to Figure 103 all help to reduce sleep disturbance particularly in apartment buildings.

Lercher et al. (2011) reported a reduced risk of hypertension in residents whose bedrooms faced a quiet courtyard, and Selander et al. (2009) reported elevated risk estimates for myocardial infarction for residents with bedrooms affected by noise. Figure 103 and Figure 105 shown apartment plans that use space planning techniques such as the stacking of bedrooms on the quiet side. By putting the bedroom window in the corner and stacking the bedrooms you can optimalise the bedroom windows with access to the quiet façade. Similarly increasing the surface area as illustrated in Figure 94 allows for more habitable space to have access to the building envelope on the quiet façade.

Research suggests it is also important to locate apartment balconies and patios on the quiet side, Gidlöf-Gunnarsson et al. (2012) noted that reported noise disturbance was two times as high where a dwelling's patio or balcony faced a railway line and approximately one and a half times higher where their bedroom faced the railway line. As illustrated in Figure 101 and Figure 106 it is possible to locate balconies and patios on the quiet side with careful space planning and good design, and it is disappointing that it isn't done more widely. A key tactic for this type of apartment space planning is to make best use of the surface area on the quiet side by having the bedroom and living area both gain natural ventilation from the balcony on the quiet side.

6.4.1 SOLUTIONS USING CHIMNEY DEVICES

As mentioned in the section on natural ventilation, chimneys can be used to draw natural ventilation from the quiet side with sealed windows on the noisy side. One benefit of the chimney solution is that it may be able to draw air from a side of the building with less localised air pollution. However, one issue with this solution is that if balconies remain located on the noisy side, as illustrated in Figure 108, the problem of having balconies and patios on the noisy side remains. As previously noted, research suggests that locating balconies on the loud side can have health consequences for residents. There are many chimney type devices to act as a driver to move naturally ventilated air. As noted in the section on natural ventilation, they usually rely on air buoyancy or wind pressure. Historical examples of wind catchers from ancient Persia also cooled the air as it came through the building. Modern ventilation chimneys and windcatchers can be used to draw air while forming a barrier to the noise; however, they rely on the right weather conditions to drive the air.

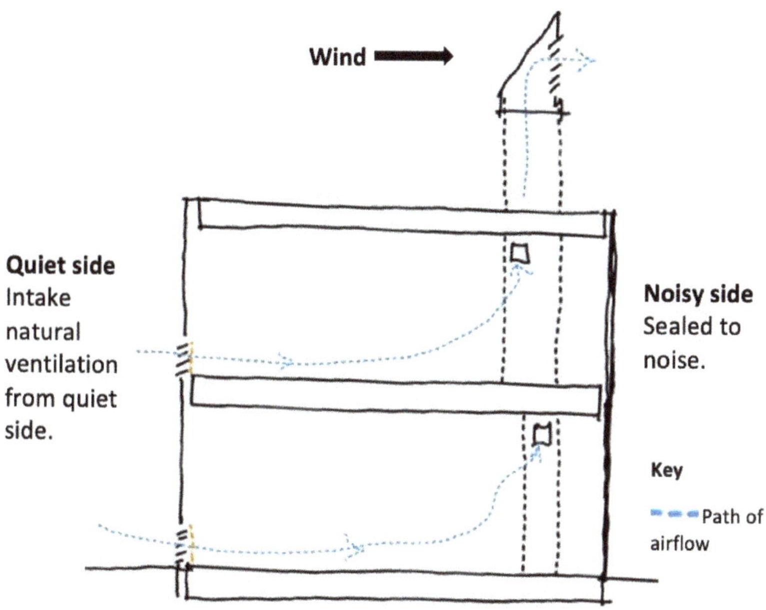

FIGURE 107 – SECTION DIAGRAM OF NOISE BARRIER BUILDING WITH VENTILATION CHIMNEY, BY J. T. COOKE; 2018, © J.T.COOKE. *NOTE: THE VENTILATION CHIMNEY USES THE WIND TO CREATE A VACUUM (NEGATIVE PRESSURE) THAT DRAWS AIR FROM THE QUIET SIDE.*

Another issue with using chimneys is they pose complex problems regarding the spread of fire; however, they can be an additional tactic to provide good human habitat when designed well. Figure 108 illustrates a project for social housing on a busy road in Sydney Australia. The building uses a plenum and chimneys to draw air from the quiet side; however, the primary balconies and patios are still on the noisy side.

Figure 108 – Public Housing 163-185 Parramatta Road, using a chimney and plenum to draw air from the quiet side; Designed by PTW Architects; Photo by J. T. Cooke; © J.T.Cooke. *Note: This tactic has the benefit of drawing fresh less polluted air from the quiet side; however, one downside is that balconies/outside space are still subject to noise.*

Readers can inspect the context of the project at the following coordinates:

33°53'14.9"S 151°10'24.3"E

and the following link:

openstreetmap.org

CHAPTER 7

SEPARATION

Tactics and Solutions Primarily for Suburban, Rural and Settlement Periphery Context

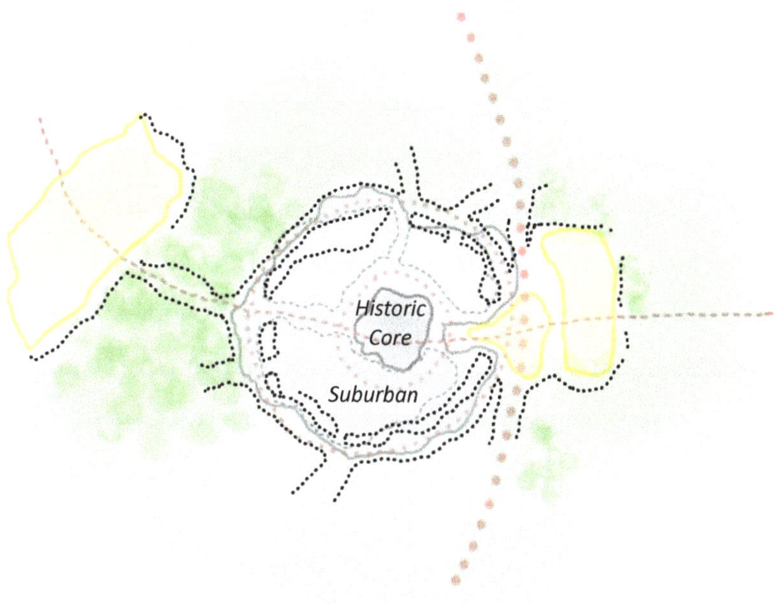

FIGURE 109 – CARICATURE OF A SETTLEMENT HIGHLIGHTING AREAS WHERE SEPARATION FROM A NOISE SOURCE IS A LIKELY TACTIC; BY J.T. COOKE; JUNE 2019, © J.T.COOKE. *NOTE: THIS IS A CARICATURE OF A SETTLEMENT, AND THE PLACES WHERE SEPARATION FROM A NOISE SOURCE IS GENERALLY USED. LOCATIONS WHERE LAND VALUES MAKE SEPARATION A VIABLE OPTION CAN INCLUDE: NEAR RING-ROADS, BUSY ROADS, AIRPORTS, INDUSTRIAL NOISE SOURCES.*

Some land uses must be separated from a population because of the negative impacts on public health. For less noxious land uses and dependent on the position of the development in the Settlement and

Habitation Transect, separation may be an economical and effective solution. In rural areas, separation alone may be an option, however in most areas, separation is usually in conjunction with some of the other solutions, such as a barrier, building layout on the site, and absorption. The effectiveness of separation depends in part on the composition and source of the sound. For a point source (e.g., a factory) an assumption of 6 dB reduction per doubling of distance or 3dB reduction per doubling of distance for linear sources (road), is often assumed (U.S Department of Transportation; Federal Highway Administration, 2017). Therefore, using separation alone as a tactic to mitigate noise can require vast separation distances that is only plausible in rural areas in the Settlement and Habitation Transect. The noise reduction is also affected by the geographic composition of the separation area, and if it is hard or soft ground. Have you ever experienced that when standing in snow covered ground, or long grass, or a ploughed field that sound behaves differently? The ground surface changes the behaviour of sound waves making it sound different to us. A fresh blanket of powdery snow absorbs sound energy making everything seem quieter. These natural phenomena can be the inspiration for design innovation. Similarly, furrows of ploughed fields diffuse and absorb sound energy and can make things seem quieter. As illustrated in Figure 111 and Figure 112 the phenomenon of diffuse surface noise propagation was used for the inspiration for Buitenschot Park project, near Schiphol airport. Conversely, a separation distance composed of hard surfaces, such as concrete runways and buildings, will tend to reflect and reverberate sound around and potentially increase noise levels.

7.1 SEPARATION OF SETTLEMENTS FROM SOURCES OF NOISE

A strategic planning approach

Land-use planning has traditionally sought to separate noise-generating land-uses from settlements to protect public health, and at strategic planning and regional level planning, separation of noise-generating activity from a settlement can be a useful tactic. With the Settlement and Habitation Transect in Table 2 and Figure 11 in mind, the separation of noise generating and noxious uses from a settlement creates an ancillary, or satellite area. The planning logic for including separation between a noisy use from a settlement is that the separation allows both land-uses to operate without impinging on the other. Separation abates public health concerns over proximity to the noxious use, and the noxious use can continue without nuisance complaints from the public. Transport links usually connect the ancillary areas to the main settlement and consequentially increases transport volumes and noise on these corridors. However, it is beneficial that particular noisy activities such as airports and industry are located away from human settlements and connected by public transport. Locating significant sources of noise away from settlements has obvious benefits. Still, government agencies must coordinate planning, transport planning, and design, and authorities should include efficient public transport links to take people to and from the 'ancillary area'. When designing and planning large pieces of regional and national transport infrastructure there should be close coordination between strategic planning, transport planning, and the design of new settlements, and new transport corridors through the existing settlement. Controls must mitigate any noise issues caused by the incorporation of the transport links on the current settlement, and designers and planners could use many of the techniques presented in this book.

7.2 SETTLEMENTS ANCILLARY AREAS

Examples of airports separated from settlements

Since the dawn of aviation, there has been tension between locating airports close enough to the main population to be accessible and far enough away so as not to cause nuisance noise. Because of the noise generated, airports are often a source of noise that is separated from a settlement, and some examples include the following:

7.2.1 STOCKHOLM ARLANDA AIRPORT EXAMPLE

Arlanda Airport is Stockholm's main airport and is located approximately 42km from Central Stockholm, and almost equidistant to Uppsalla. The large separation distance from the population centre of Stockholm seeks to ameliorate any aircraft noise for the majority of the population. There are rail, bus and road links to the airport from Stockholm, but the journey takes about an hour.

Readers can inspect the context of the airport at the following coordinates:

59°38'47.8"N 17°55'17.3"E

and the following address:

Stockholm Arlanda Airport, Stockholm, Sweden.

and the following links:

openstreetmap.org

7.2.2 COPENHAGEN AIRPORT, KASTRUP EXAMPLE

Copenhagen Airport, Kastrup is Copenhagen's main airport and is located approximately 8.3 km from the centre of Copenhagen and is located adjacent to the residential suburbs of Maglebylille and Dragør. There are good rail, bus and road links to Copenhagen and travel times are approximately 15 minutes. Areas of Maglebylille were compulsorily purchased by the government in the 1960's to make way for the airport. However, the residential areas of both the main settlement of Copenhagen and the suburbs of Maglebylille and Dragør are far closer to the airport when compared to Arlanda Airport. Readers can examine the context at the link below. The main take-off and landing path of the aircraft is over the sea and there is little inhabitation directly under these routes. There is also some commercial uses between Maglebylille and the airport that may act as a barrier for aircraft ground noise for the residential areas.

Readers can inspect the context of the airport at the following coordinates:

55°37'35.6"N 12°38'17.0"E

and the following address:

Lufthavnsboulevarden 6, 2770 Kastrup, Denmark

and the following links:

openstreetmap.org

7.2.3 HELSINKI-VANTAA AIRPORT EXAMPLE

Helsinki Airport is Helsinki's main airport and is located approximately 19km from the main settlement of Helsinki, in the neighbouring city of Vantaa. The main east-west runways have take-off and landing routes located primarily over rural and industrial areas, however; the north-south runway passes over residential areas and residents are affected by aircraft noise. As with the other examples there are separation distances and some barrier commercial uses between the airport and the residential area that help mitigate aircraft ground noise. There are good rail, bus and road links to Helsinki and travel times are approximately 20 to 40 minutes.

Readers can inspect the context of the airport at the following coordinates:

60°18'47.7"N 24°58'38.3"E

and the following address:

Helsinki Airport

01531 Vantaa,

Finland

and the following links:

openstreetmap.org

7.2.4 SYDNEY KINGSFORD SMITH AIRPORT EXAMPLE

Sydney Kingsford Airport is Sydney's main airport and includes both domestic and international flights. The airport is approximately 8.4 km from the centre of Sydney and has good rail, bus and road links, and travel times are about 15mins. The main runways are in a north-south direction, and take-off and landing routes are over the sea to the south but are over residential and industrial areas to the north, as in the example of Copenhagen Airport having the sea as a separation area is a useful tactic. The residential areas of Sydenham, St Peters and Marrickville to the north are affected by aircraft noise and properties were compulsory purchased in the past to create a separation distance for the airport. There are shorter runways oriented east-west, for take-off and landing routes and the suburbs of Eastlakes and Banksia are affected by aircraft noise. As with the other examples, a barrier of industrial and commercial land-use in some areas help mitigate aircraft ground noise to the nearby residential areas.

Readers can inspect the context of the project at the following coordinates:

33°56'13.7"S 151°10'27.2"E

and the following address:

Sydney Airport

Sydney NSW 2020

and the following links:

openstreetmap.org

7.2.5 LEONARDO DA VINCI AIRPORT EXAMPLE

Leonardo da Vinci International Airport is Rome's main airport and is located approximately 35 km from the centre of Rome. The main runways are in a north-south direction, with smaller runways in an east-west orientation. The most westerly north-south runway has take-off and landing routes that affect Fiumicino to the south of the airport, but the other runways are primarily over rural areas. There are good rail, bus and road links to Rome and travel times are approximately 30 minutes.

Readers can inspect the context of the airport at the following coordinates:

41°47'48.1"N 12°14'55.3"E

and the following address:

Via dell' Aeroporto di Fiumicino,

00054 Fiumicino RM,

Italy

and the following links:

openstreetmap.org

In the examples presented the reader may have noticed while analysing the strategic and regional planning of airports and their settlements that varying amounts of separation from noise sensitive uses are used. A distance of approximately 30 km is the average separation for the examples shown and there is an optimal distance to be found that mitigates noise and provides reasonable travel times to the centre of the settlement. There have been advances in aircraft design to reduce the noise generated by aeroplanes; however, problems caused by aircraft noise are an ongoing issue. Readers can find some further helpful information on aircraft noise at the link below:

nats.aero

7.3 NOISE, SEPARATION & SETTLEMENT GROWTH

Where can settlements grow?

It is prudent to locate noisy land-uses away from the main settlements. As our communities grow, and there is development pressure for scare land our settlements invariably seek to grow into the separation distance provided from noxious and noisy land-uses. In the examples illustrated above, separation distances were often rural areas or a combination of a barrier of the commercial regions for aircraft ground noise and separation. Often in densely populated areas, there is a high demand for land, and sometimes another less sensitive land-use can be used as a barrier in conjunction with a separation zone. The section on *Noise barrier structures* illustrated the example of The Wall Shopping Centre in the Netherlands that created a noise barrier and allowed further housing to be built behind the barrier. The less sensitive land-use is often a building that is already going to be sealed and air-conditioned or could be an open space, such as a park. The use of noise reduction techniques can, to a degree, reduce the amount of separation required and increase the amount of land available for a settlement to grow.

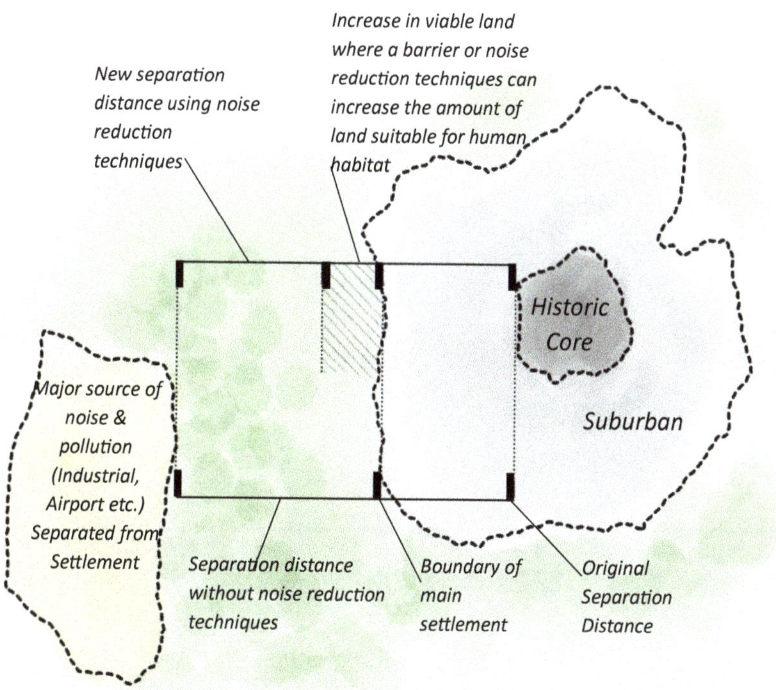

FIGURE 110 - CARICATURE OF A SETTLEMENT & DIAGRAM OF THE POTENTIAL INCREASE IN THE AREA OF LAND FIT FOR HUMAN HABITATION WITH THE USE OF NOISE REDUCTION TECHNIQUES; BY J.T. COOKE; JUNE 2019, © J.T.COOKE. *NOTE: THE USE OF NOISE REDUCTION TECHNIQUES CAN INCREASE THE LAND VALUES IN THE NEWLY CREATED QUIETER LAND BY MAKING IT VIABLE FOR HUMAN HABITATION AND HOUSING. THIS IMAGE IS A CARICATURE OF A SETTLEMENT, AND IN REALITY, THE INCREASE IN LAND FEASIBLE FOR HUMAN HABITATION WITH THE USE OF NOISE REDUCTION TECHNIQUES CAN OCCUR WITHIN EXISTING SETTLEMENTS, ON RING ROADS, OR ADJACENT TO OTHER NOISE SOURCES. THE USE OF NOISE REDUCTION TECHNIQUES CAN INCREASE THE VALUE OF LAND IN THE NEWLY CREATED QUIET AREA.*

As illustrated in Figure 110, the use of noise reduction techniques can also increase the land values in the newly created quieter land by making it viable for human habitation and housing, or making an area beyond the barrier suitable, as in The Wall Shopping Centre example. Figure 110 is a caricature of a settlement, and in reality, the increase in land feasible for human habitation with the use of noise reduction techniques can also occur within existing settlements, on ring roads, or adjacent to other noise sources. The use of noise reduction techniques can increase land value in the newly created quiet area by making it suitable for human habitation.

FIGURE 111 - INTERNATIONAL AWARD-WINNING DESIGN FOR ACOUSTIC ATTENUATION LANDFORM AND RECREATION PARK; BUITENSCHOT PARK, NEAR SCHIPHOL AIRPORT, COURTESY OF H+N+S LANDSCAPE ARCHITECTS, HNSLAND.NL; IMAGE COURTESY OF © YOUR CAPTAIN LUCHTFOTOGRAFIE; 2014; WEB; DECEMBER 2017. *NOTE: THE LANDFORM MAKES SHAPES LIKE A PLOUGHED FIELD TO SCATTER AND DIFFUSE SOUND ENERGY.*

The separation distance can also be a multi-use space, and continue to provide open space as separation, albeit a reduced amount of separation distance. Buitenschot Park is an innovative example of a reduced separation distance, park, and noise reduction landform. Designed by H+N+S landscape architects in collaboration with artist Paul De Kort, TNO and Witteveen en Bos Consultant, the park mitigates ground noise generated by aeroplanes on the runway at Schiphol Airport (see Figure 111 and Figure 112). The unique acoustic landform of the park enhances the performance of the separation area.

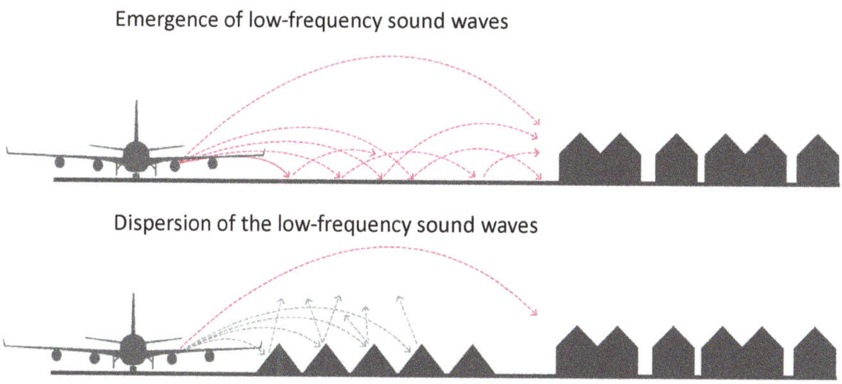

Figure 112 - International award-winning design for acoustic attenuation landform and recreation park; Buitenschot Park, near Schiphol Airport, courtesy of H+N+S Landscape Architects, hnsland.nl; 2014; web; December 2017 © H+N+S. *Note: The low frequency sound waves are dispersed by the ploughed furrow like shapes of Buitenschot park landforms. The noise reaching the housing is reduced.*

The noise reduction park is designed specifically to mitigate ground noise generated by aircraft on the runways, and is not targeted at reducing noise from airborne aircraft. The Buitenschot Park project proves that, depending on the noise source and with innovative design, separation solutions can perform multiple functions and create exciting spaces in a separation zone. It is unfortunate that there aren't more examples like Buitenschot Park, where acousticians, landscape architects, architects, and artists work together to make something beautiful and practical.

Readers can inspect the context of the project at the following coordinates:

52°19'10.4"N 4°41'09.3"E

and the following address:

Landartpark Buitenschot

Hoofddorp

Netherlands

and the following links:

Google.com/maps

CONCLUSION

Noise is a pervasive public health issue, and climate change is a risk to our world, so proposing air-conditioning where human habitation is proximate to noise should not be the default noise mitigation solution. This book has illustrated many practical design solutions that can mitigate noise and provide passive ventilative cooling in various contexts, and seeks to reinforce that air-conditioning need not be the default solution.

This book discussed noise in human settlements and presented a suite of design solutions within the Settlement & Habitation Transect framework, which categorises context to assist in identifying where a specific noise mitigation design solution may be most appropriate.

To help accomplish the WHO guiding principles identified in the *Environmental Noise Guidelines for the European Region* an increased knowledge of the passive natural ventilation and noise mitigation solutions is needed among built-environment professionals to promote design and planning that reduces noise exposure of the inhabitants of a settlement.

The Settlement & Habitation Transect could be a shared high-level structuring tool to encourage collaboration and understanding between everyone involved in the creation and management of settlements, and the communities who inhabit them.

Hopefully, this book will help promote greater awareness of the design of passive natural ventilation with passive noise mitigation, and the potential solutions that respond to the structure of a settlement and create suitable human habitat.

POSTSCRIPT

Hopefully this book has been useful and interesting. Further information, tutorials, and resources are available on the Kuu-Media website at www.kuu-media.com. An accompanying video tutorial series about designing for noise is available on the Kuu Media website.

Please subscribe to the Media YouTube channel for free videos.

Kuu-Media.com

YouTube Channel

APPENDIX

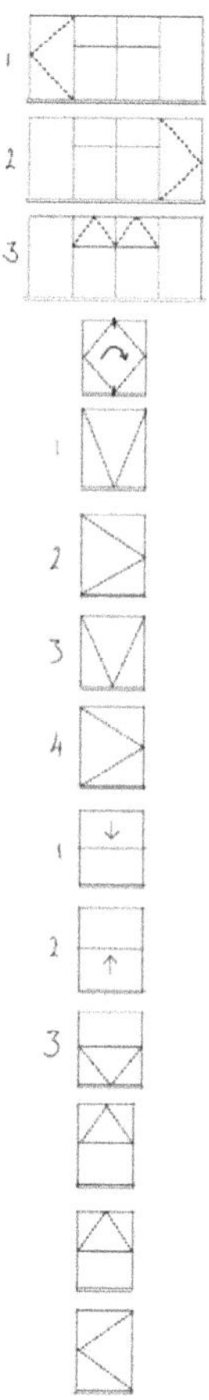

	Comparative Level Difference (dBA)			
	D_A road	D_A rail	D_A air	D_A music
A 1	17	17	18	16
A 2	17	17	18	16
A 3	14	14	16	16
B	12	12	14	15
C 1	16	16	17	19
C 2	17	17	19	20
C 3	16	16	18	19
C 4	15	15	17	18
D 1	16	16	18	18
D 2	14	14	16	17
D 3	16	16	18	18
E	17	17	18	18
F	18	18	18	18
G	15	15	15	17

dBA level difference for different source characteristics (50k mm2 open windows)

THE dB(A) LEVEL DIFFERENCE FOR DIFFERENT SOURCE CHARACTERISTICS AND VARIOUS TYPES OF WINDOWS TESTED. ADAPTED FROM; OPEN/CLOSED WINDOW RESEARCH SOUND INSULATION THROUGH VENTILATED DOMESTIC WINDOWS (P88) BY WATER-FULLER ET. AL., 2007, RETRIEVED FROM; HTTPS://RMP.BIZ/PORTFOLIO-POSTS/DEFRA-OPEN-WINDOWS-PROJECT/. © CROWN COPYRIGHT, 2007, THIS IMAGE IS LICENCED UNDER OPEN GOVERNMENT LICENCE 3.0 HTTP://WWW.NATIONALARCHIVES.GOV.UK/DOC/OPEN-GOVERNMENT-LICENCE/VERSION/3/ [5]

ABBREVIATIONS

WHO – World Health Organization

UN – United Nations

dB – Decibel

dB(A) – A-weighted decibel

HVAC - Heating Ventilation and Air Conditioning systems

AAPV - Acoustic attenuated passive ventilation

GLOSSARY OF TERMS AND DEFINITIONS

Acoustic Environment – The physical phenomenon of sound in an environment. (ISO 12913-1 standard)

Acoustic Context – The physical phenomenon of sound in specific contexts.

A-weighted - a weighting applied to a measured sound pressure level to account for the human perception of loudness

Negative pressure – A negative pressure is a vacuum. The phrase negative pressure is commonly used in the design and building industry.

Noise – An unpleasant sound that causes disturbance.

Over Station Development – Developments built directly over train stations. Examples are in London, New York, Chicago, Sydney

Settlement and Habitation Transect – The term Settlement and Habitation Transect is used rather than Urban Transect as it is felt a better description of what the transect is structuring.

Slot ventilation - a small opening in the window frame to provide trickle ventilation

Sound Profile - The sound and noise profiles are the varying pressure levels and frequency that our brains recognise as the characteristics of sounds. Sound & Noise profiles are specific, and the attenuation required needs to remove or deaden unwanted sound also need to be specific to the sound profile. Some sounds have more bass elements, some more high pitch, and sound mitigation needs to target the frequencies of the type of sound you are trying to diminish. A Sound Spectrograph using a Fournier Transform can describe and give its profile in a sound spectrograph, like a voiceprint.

Soundscape – Is the human perceptual construct of a sound environment. (ISO 12913-1 standard)

Streetwall - the enclosure walls around a street that the buildings adjacent to the street create (see Calle de Atocha in Madrid, and Plaza Major, note the walls of the street or square are created by the external wall of the buildings and give the place its sense of enclosure like an external room without a ceiling.

Tabula Rasa - a blank canvas.

Terroir - This term includes all the subtle variations in the land, the geography that in different proportions go to make a settlement unique, and like a good wine time is vital in the formation of a settlement.

Yield - the annual rent or sales return that an investor receives in the form of income or lump sum, compared to the amount of money expended to buy a site or property.

BIBLIOGRAPHY

Öhrström, E., Skånberg, A., Svensson, H., & Gidlöf-Gunnarsson, A. (2006, August 8). Effects of road traffic noise and the benefit of access to quietness. *Journal of Sound and Vibration*, 40-59.

Basner, M., & McGuire, S. (2018). WHO Environmental Noise Guidelines for the European Region: A Systematic Review on Environmental Noise and Effects on Sleep. *International Journal of environmental research and public health*, 15-3519. doi:10.3390/ijerph15030519

Bluhm, G., Nordling, E., & Berglind, N. (2004, 06 12). *Road traffic noise and annoyance—An increasing environmental health problem.* Retrieved 10 15, 2018, from Noise and Health: http://www.noiseandhealth.org/article.asp?issn=1463-1741;year=2004;volume=6;issue=24;spage=43;epage=49;aulast=Bluhm

Bruce, N., & Davies, W. (2014). *The effects of expectation on the perception of soundscapes.* Manchester: University of Salford, Manchester. Retrieved from http://usir.salford.ac.uk/31873/

Conference of European Directors of Roads. (2019, December 2). *On-Air.* Retrieved from reports and results: www.on-air.no/oa_uploads/ON_AIR_Guidancebook_2015-12-22.pdf

de Kluizenaar, Y., Janssen, S. A., Vos, H., Salomons, E. M., Zhou, H., & van den Berg, F. (2013). Road Traffic Noise and Annoyance: A Quantification of the Effect of Quiet Side Exposure at Dwellings. *International Journal of Environmental Health*, 1-9.

Department for Environment, Food and Rural Affairs. (2018, June 28). *Noise pollution: economic analysis*. Retrieved 08 23, 2018, from www.gov.uk: https://www.gov.uk/guidance/noise-pollution-economic-analysis

Department for Town Planning and Building Regulations. (2012). *Peace at last; new dwellings on the middle ring road.* Munich: State Capital Munich.

Duany Plater-Zyberk & Company. (2019, May 30). *Center for Applied Transect Studies.* Retrieved January 30, 2019

Fastl, H., & Zwicker, E. (2007). *Psychoacoustics; Facts and Models.* Berlin: Springer-verlag. doi:https://doi.org/10.1007/978-3-540-68888-4_2

Fenech, B. (2017, April 26). *Noise-related sleep disturbance - a brief overview of the scientific evidence.* Retrieved April 23, 2018, from Association of Acoustic and Noise Consultants: http://www.association-of-noise-consultants.co.uk/conference-2017-presentations/

Gidlöf-Gunnarsson, A., Ögren, M., & Jer, T. (2012, 08 12). *Railway noise annoyance and the importance of number of trains, ground vibration, and building situational factors.* Retrieved 11 16, 2018, from Noise and Health: http://www.noiseandhealth.org/article.asp?issn=1463-1741;year=2012;volume=14;issue=59;spage=190;epage=201;aulast=Gidl%F6f%2DGunnarsson

Hilpert, D. J. (1845). A Dictionary of the English and German, and the German and English Language. *2, Part 2*, 173. Vienna, Austria: Th. Braun. Retrieved from https://books.google.com.au/books?id=tG5eAAAAcAAJ&lpg=PA173&dq=what%20does%20Pf%C3%B6rtchen%20mean%20in%20german%3F&pg=PP9#v=onepage&q=what%20does%20Pf%C3%B6rtchen%20mean%20in%20german?&f=false

HL HO, J., & WM CHU, C. (2016). *Research and Development of Noise Mitigation Measures for Public Housing Development in Hong Kong - A Case Study of Acoustic Balcony.* Hong Kong: Hong Kong Housing Department. Retrieved from http://pub.dega-akustik.de/IN2016/data/articles/000430.pdf

Lawrence, F., & Kavage, S. (2008). Urban Planning and Public Health: A Story of Separation and Reconnection. *Journal of Public Health Management and Practice*, 214-220. Retrieved from http://

journals.lww.com/jphmp/Citation/2008/05000/Urban_Planning_and_Public_Health

Lercher, P., Botteldooren, D., Widmann, U., & Uhrner, U. (2011, 04 29). *Cardiovascular effects of environmental noise: Research in Austria.* Retrieved 10 17, 2018, from Noise and Health.

Lurcock, D., MacKenzie, R., MacKenzie, R., & Waters-Fuller, T. (2006). *Open/closed window research: sound insulation through ventilated domestic windows.* Retrieved from Edinburgh Napier University Research Output: https://www.napier.ac.uk/research-and-innovation/research-search/outputs/openclosed-window-research-sound-insulation-through-ventilated-domestic-windows

Marshall, J. H. (1931). *Mohenjo-Daro and the Indus Valley Civiization.* Hertford: Asian Educational Services; Facsimile of 1931 ed edition. Retrieved from https://archive.org/details/in.ernet.dli.2015.722/page/n51

Ministry of Urban Development and Environment. (2011). *Soundproofing with partially open windows.* Retrieved from HafenCity Hamburg: https://www.hafencity.com/upload/files/files/Laerm_Leitfaden_3_1.pdf

National Academy of Engineering of the National Academies. (2010, May 4). *Technology for a Quieter America - Metrics for Assessing Environmental Noise.* Retrieved January 31, 2019, from https://www.nap.edu/read/12928/chapter/1#ii

National Building Research Institute; Aalborg University Copenhagen. (2013, 12 02). *SBI instruction 244 Soundproofing of the climate screen.* Retrieved from National Building Research Institute; Aalborg University Copenhagen: https://sbi.dk/anvisninger/Pages/244-Lydisolering-af-klimaskaermen-1.aspx#/5-Lydisolering-af-eksisterende-bygningers-klimaskaerm/5-3-Facadevinduer

Nijs, L., Versteeg, P., & van der Voorden, M. (2004). *The combination of absorbing materials and room shapes to reduce noise levels.* Delft: Delft University of Technology.

Rasmussen, B. (2015). *Experiences with sound insulating open windows in traffic noise-exposed housing*. Copenhagen: Aalborg University.

Søndergaard, L. S., & Egedal, R. (2016). *Open windows with better sound insulation*. Hamburg: Inter Noise. Retrieved from http://assets.madebydelta.com/docs/share/Akustik/open_windows_with_better_sound_insulation_internoise_2016.pdf

Salomons, E., de Kluizenaar, Y., Botteldooren, D., Wei, W., & Ögren, M. (2013, August 23). *Project QSide*. Retrieved June 18, 2018, from QSide: http://qside.se

Selander, J., Nilsson, M., Bluhm, G., Rosenlund, M., Lindqvist, M., Nise, G., & Pershagen, G. (2009, 03 01). *Long-Term Exposure to Road Traffic Noise and Myocardial Infarction*. Retrieved 10 17, 2018, from Epidemiology: https://journals.lww.com/epidem/fulltext/2009/03000/Long_Term_Exposure_to_Road_Traffic_Noise_and.20.aspx

SONORUS Project. (2016). *Urban Sound Planning - the SONORUS project*. Gothenburg: Chalmers University of Technology. Retrieved from https://www.google.co.uk/url?sa=t&rct=j&q=&esrc=s&source=web&cd=1&cad=rja&uact=8&ved=2ahUKEwjNrILPivjaAhWrAsAKHRAiA-gQFjAAegQIABAr&url=https%3A%2F%2Fwww.researchgate.net%2Fprofile%2FFrancesco_Aletta%2Fpublication%2F308954398_URBAN_SOUND_PLANNING_-_the_SON

U.S Department of Transportation; Federal Highway Administration. (2017, September 06). *Physical Techniques to Reduce Noise*. Retrieved June 14, 2018, from The Audible Landscape: A Manual for Highway Noise: https://www.fhwa.dot.gov/Environment/noise/noise_compatible_planning/federal_approach/audible_landscape/al00.cfm

United Nations. (2018, March 29). *United Nations Secretary-General: Media*. Retrieved 02 17, 2019, from United Nations: https://www.un.org/sg/en/content/sg/press-encounter/2018-03-29/secretary-generals-press-encounter-climate-change-qa

United Nations. (2018, May 16). *United Nations: Department of Economic and Social Affairs*. Retrieved January 30, 2019

Van Renterghem, T., & Botteldooren, D. (2012). Focused Study on the Quiet Side Effect in Dwellings Highly Exposed to Road Traffic Noise. *International Journal of Environmental Research and Public Health*, 4293-4308. Retrieved from https://www.mdpi.com/1660-4601/9/12/4292

Wilkey, E. (1839, January 1). *Wanderings in Germany: With Moonlight Walks on the Banks of the Elbe, the Danube, the Neckar, and the Rhine*. Retrieved from Google Books: https://books.google.com.au/books?id=VsZDAQAAMAAJ&lpg=PR1&ots=gFNDh9EUhC&dq=Wanderings%20in%20Germany%2C%20With%20Moonlight%20Walks%20on%20the%20Banks%20of%20the%20Elbe%2C%20the%20Danube%2C%20the%20Neckar%2C%20and%20the%20Rhine&pg=PR1#v=onepage&q=Wanderin

World Health Organization. (2018, June 25). *Burden of Disease from Environmental Noise; Quantification of healthy life years lost in Europe*. Retrieved 08 23, 2018, from World Health Organisation Europe: http://www.euro.who.int/__data/assets/pdf_file/0008/136466/e94888.pdf

World Health Organization. (2018, September 10). *Environmental Noise Guidelines for the European Region*. Retrieved September 15, 2018, from World Health Organization: Europe: http://www.euro.who.int/en/health-topics/environment-and-health/noise/publications/2018/environmental-noise-guidelines-for-the-european-region-2018

Yeung, M. (2016). *Adopting Specially Designed Balconies to Achieve Substantial Noise Reduction for Residential Buildings*. Hamburg: Internoise.

ATTRIBUTION STATEMENTS

1 *'There is biologic plausibility that chronic night-time exposure to relevant levels of noise can contribute to negative health consequences like a cardiovascular disease'*

Reprinted from; "WHO Environmental WHO Environmental Noise Guidelines for the European Region: A Systematic Review on Environmental Noise and Effects on Sleep" by Basner, M., & McGuire, S. (2018). International Journal of Environmental Research and Public Health, 15(3), 519. MDPI AG. Retrieved from http://dx.doi.org/10.3390/ijerph15030519., licensed under CC BY 4.0 https://creativecommons.org/licenses/by/4.0/

2 *'...disability-adjusted life years' (DALYs) from noise-related issues, such as: ischemic heart diseases (61,000 years), cognitive impairment of children (45,000 years), sleep disturbance (903,000 years), tinnitus (22,000 years), annoyance (654,000 years)'.*

Reprinted from; World Health Organization; Burden of Disease from Environmental Noise; Quantification of healthy life years lost in Europe. By World Health Organisation Europe; Conclusions; Estimated DALYs for western European countries; Page No 101.; Retrieved 08 23, 2018, from http://www.euro.who.int/__data/assets/pdf_file/0008/136466/e94888.pdf © WHO Copyright (2011); Reprinted with the permission of the World Health Organization.

3 *'Rotation of source incidence away from the normal, within a non-diffuse acoustic environment, is found to consistently improve the resulting open window façade insulation'*

Reprinted from; Lurcock, D., MacKenzie, R., MacKenzie, R., & Waters-Fuller, T. (2007, April,p.5). *NANR116: Open/closed window research: sound insulation through ventilated domestic windows.* Retrieved 2019, from DEFRA Open Windows Project; https://rmp.biz/portfolio-posts/defra-open-windows-project/. © Crown copyright 2007, This text is licenced under Open Government Licence 3.0 http://www.nationalarchives.gov.uk/doc/open-government-licence/version/3/

4 *'...free open area of 0.05 m2 the measured insulation ratings ranged between DW 14 and 20dB, and translated to dBA level differences of:*

Road traffic noise 12 – 18 dBA

Railway noise 12 – 18 dBA

Aircraft noise 14 – 19 dBA

Amplified music 15 – 20 dBA

The study found that there was little variation because of the window style, nor the material. The study also investigated closed windows with a 4000mm2 slot and found that the ventilator reduced the performance of the window by 6dB when closed and 11dB when open; therefore, a considerable decrease in performance.

5 And Appendix table; The dB(A) level difference for different source

characteristics and various types of windows tested

Adapted from; Lurcock, D., MacKenzie, R., MacKenzie, R., & Waters-Fuller, T. (2007, April,p.5). *NANR116: Open/closed window research: sound insulation through ventilated domestic windows*. Retrieved 2019, from DEFRA Open Windows Project; https://rmp.biz/portfolio-posts/defra-open-windows-project/. © Crown copyright 2007, This text is licenced under Open Government Licence 3.0 http://www.nationalarchives.gov.uk/doc/open-government-licence/version/3/

6 *'the most systemic threat to humankind'*.

From United Nations Secretary-General's press encounter on climate change [with Q&A], by António Guterres. ©United Nations 2018. Reprinted with the permission of the United Nations. Retrieved 02 17, 2019, from United Nations Secretary-General: https://www.un.org/sg/en/content/sg/press-encounter/2018-03-29/secretary-generals-press-encounter-climate-change-qa

7 68% of the world population will live in urban areas by 2050

From United Nations: Department of Economic and Social Affairs; News, by United Nations. ©United Nations 2018. Reprinted with the permission of the United Nations. Retrieved January 30, 2019, from United Nations: https://www.un.org/development/desa/en/news/population/2018-revision-of-world-urbanization-prospects.html

8 *'..[I]in most cases, the maximum noise reduction that can be achieved by a barrier is 20 dB(A) for thin walls and 23 dB(A) for berms'*

and

9 *'Studies have shown that a T-profile top barrier provides insertion losses comparable to a conventional top barrier when the difference in their heights is equal to the width of the T-profile top. When the two barriers are the same height, the T-profile top barrier has been shown to provide an additional 2.5 dB(A) insertion loss over the conventional top barrier'*

From U.S Department of Transportation; Federal Highway Administration. (2017, September 06). *Physical Techniques to Reduce Noise*. Retrieved June 14, 2018, from The Audible Landscape: A Manual for Highway Noise: https://www.fhwa.dot.gov/Environment/noise/noise_compatible_planning/federal_approach/audible_landscape/al00.cfm. This is a work of the U.S. Government and is not subject to copyright protection in the United States. Foreign copyrights may apply.

10 *'Context, expectations, and people's experiences and circumstances all affect their responses. Hence, levels of community response (e.g., annoyance) may vary from community to community, just as individual responses vary from person to person, even if noise levels do not change. However, one consistent finding has been that changes in noise exposure do affect individual and community responses and that increases in man-made noise usually have a negative impact.'*

From *Technology for a Quieter America -Chapter 3; Metrics for Assessing Environmental Noise, para 3, p19*. © National Academy of Sciences May 2010, Reprinted with the permission of the National Academy of Science. Retrieved January 31, 2019, from https://www.nap.edu/read/12928/chapter/1#ii

LIST OF FIGURES

TABLES

INDEX

A

B

C

D

E

F

G

H

I

L

M

N

O

P

Q

R

S

T

U

W

Y

ABOUT THE AUTHOR

J. T. Cooke, B.A. (Hons), MArch, DipArch, M.A. RIBA

J.T. Cooke is an Architect, Urban Planner and Urban Designer with over twenty years of experience in building design, urban design, place planning and strategic planning in the U.K., Ireland, and Australia. He has extensive experience in architectural practice and planning, and has been an expert witness in court. He operates STÅ consultancy based in Sydney, Australia, and is involved in education.

J. T. Cooke studied at Kingston University, Edinburgh University, and Cardiff University and has taught at the University of Sydney, University of Technology Sydney (UTS), and the University of New South Wales (UNSW). He has also been a guest critic at the University of West of England, the University of New South Wales, University Technology Sydney, and the University of Sydney, and is a guest speaker at Cardiff University.

J. T. Cooke puts people at the centre of design and planning. He has always been passionate about how design and planning decisions affect peoples lives. He believes that what architects & planners draw and legislate for becomes our tomorrow, and we must improve the quality of life for everyone and tackle climate change to have a tomorrow.

www.ingramcontent.com/pod-product-compliance
Ingram Content Group UK Ltd.
Pitfield, Milton Keynes, MK11 3LW, UK
UKHW020657041026
15590UKWH00001B/329

9 781739 869205